DATE DUE			
GAYLORD M-2			PRINTED IN U.S.A.

CHEMICAL ANALYSIS

Vol. 1. **The Analytical Chemistry of Industrial Poisons, Hazards, and Solvents.** *Second Edition.* By Morris B. Jacobs

Vol. 2. **Chromatographic Adsorption Analysis.** By Harold H. Strain (*out of print*)

Vol. 3. **Colorimetric Determination of Traces of Metals.** *Third Edition.* By E. B. Sandell

Vol. 4. **Organic Reagents Used in Gravimetric and Volumetric Analysis.** By John F. Flagg (*out of print*)

Vol. 5. **Aquametry: Application of the Karl Fischer Reagent to Quantitative Analyses Involving Water.** By John Mitchell, Jr. and Donald Milton Smith (*temporarily out of print*)

Vol. 6. **Analysis of Insecticides and Acaricides.** By Francis A. Gunther and Roger C. Blinn (*out of print*)

Vol. 7. **Chemical Analysis of Industrial Solvents.** By Morris B. Jacobs and Leopold Scheflan

Vol. 8. **Colorimetric Determination of Nonmetals.** Edited by David F. Boltz

Vol. 9. **Analytical Chemistry of Titanium Metals and Compounds.** By Maurice Codell

Vol. 10. **The Chemical Analysis of Air Pollutants.** By Morris B. Jacobs

Vol. 11. **X-Ray Spectrochemical Analysis.** *Second Edition.* By L. S. Birks

Vol. 12. **Systematic Analysis of Surface-Active Agents.** By Milton J. Rosen and Henry A. Goldsmith

Vol. 13. **Alternating Current Polarography and Tensammetry.** By B. Breyer and H. H. Bauer

Vol. 14. **Flame Photometry.** By R. Herrmann and J. Alkemade

Vol. 15. **The Titration of Organic Compounds.** (*In two Parts*) By M. R. F. Ashworth

Vol. 16. **Complexation in Analytical Chemistry: A Guide for the Critical Selection of Analytical Methods Based on Complexation Reactions.** By Anders Ringbom

Vol. 17. **Electron Probe Microanalysis.** By L. S. Birks

Vol. 18. **Organic Complexing Reagents: Structure, Behavior, and Application to Inorganic Analysis.** By D. D. Perrin

Vol. 19. **Thermal Methods of Analysis.** By Wesley Wm. Wendlandt

Vol. 20. **Amperometric Titrations.** By John T. Stock

Vol. 21. **Reflectance Spectroscopy.** By Wesley Wm. Wendlandt and Harry G. Hecht

Vol. 22. **The Analytical Toxicology of Industrial Inorganic Poisons.** By the late Morris B. Jacobs

Vol. 23. **The Formation and Properties of Precipitates.** By Alan G. Walton

Vol. 24. **Kinetics in Analytical Chemistry.** By Harry B. Mark, Jr. and Garry A. Rechnitz

Vol. 25. **Atomic Absorption Spectroscopy.** By Walter Slavin

Vol. 26. **Characterization of Organometallic Compounds.** (*In two Parts*) Edited by Minoru Tsutsui

Vol. 27. **Rock and Mineral Analysis.** By John A. Maxwell

Vol. 28. **Analytical Chemistry of Nitrogen and Its Compounds.** (*In two parts*) Edited by C. A. Streuli and P. R. Averell

Vol. 29. **Analytical Chemistry of Sulfur and Its Compounds.** (*In two parts*) By J. H. Karchmer

Other volumes in preparation

CHEMICAL ANALYSIS

A SERIES OF MONOGRAPHS ON ANALYTICAL CHEMISTRY AND ITS APPLICATIONS

VOLUME 11

INTERSCIENCE PUBLISHERS

A division of John Wiley & Sons, New York/London/Sydney/Toronto

X-Ray Spectrochemical Analysis

SECOND EDITION

L. S. BIRKS

X-Ray Optics Branch
U.S. Naval Research Laboratory
Washington, D.C.

1969

INTERSCIENCE PUBLISHERS

A division of John Wiley & Sons, New York/London/Sydney/Toronto

The paper used in this book has pH of 6.5 or higher. It has been used because the best information now available indicates that this will contribute to its longevity.

1 2 3 4 5 6 7 8 9 10

Library of Congress Catalog Card Number 71-79144

SBN 471 07525 6

PRINTED IN THE UNITED STATES OF AMERICA

To my family for their forbearance
and to my colleagues for their help and advice.

PREFACE TO SECOND EDITION

In the ten years since the publication of the first edition, there have been more changes and improvements in the capabilities of X-ray spectrochemical analysis than in all the previous history of the subject. For instance, availability of long-spacing synthetic crystals and stearate layers has extended the range of elements downward to boron. Solid-state detectors have improved energy dispersion by a factor of 3 for the middle-range elements. Mathematical methods for quantitative analysis have advanced, with the aid of computers, to the point where one can truly begin to eliminate the need for comparison standards. X-ray tubes with higher wattage, thinner windows, and a wider variety of targets have pushed the limit of detectability to the parts per million range for many elements.

The objective and approach of this edition remains the same as for the first edition, to point out the underlying principles and practices which govern the generation, dispersion, and detection of characteristic radiation and to show how the data are interpreted to achieve quantitative analysis. General types of applications are presented as examples of the kinds of analyses which may be performed. The chapter on electron probe microanalysis has been revised considerably to conform to the present availability of commercial instruments which require less understanding of electron optics and details of construction by operators; specialized books on the electron probe go into detail on these subjects.

I wish to thank my colleague, John V. Gilfrich, who proofread the manuscript and made many helpful suggestions. I also wish to thank the other members of the X-Ray Optics Branch who have contributed to the advancement of X-ray spectrochemical analysis over the years and helped make many of the advances discussed here possible.

L. S. Birks

April 1969

PREFACE TO FIRST EDITION

It has been over 25 years since G. von Hevesy's excellent book *Chemical Analysis by X-Rays and its Applications* was published and there has been no other book on the subject during those years. It is not time alone, however, but changes in instrumentation and technique that indicate the need for a new book on X-ray spectrochemical analysis. I have tried to bring the subject up to date for the scientist who is interested in X-ray spectrochemistry as a research tool and also to present the material in a way that will be useful to those persons who are only interested in knowing enough about the method to be able to use it judiciously for routine analysis.

With the knowledge that the majority of readers will not start in and read from cover to cover, I have attempted to arrange the material on excitation, dispersion, detection, and quantitative analysis so that each aspect stands alone. The chapter on applications is intended to point out general types of applications rather than give specific information on a particular subject because there is still rapid change in X-ray spectroscopy, and specific information that is valid today may well be misleading in a few years. Chapter 7 on the electron probe microanalyzer recognizes this new tool as being enough different from fluorescent X-ray spectroscopy to require separate treatment. This is not only because the instrumentation is of a rather different type, but also because the applications will frequently not overlap those of fluorescent analysis.

I wish to express my thanks to E. J. Brooks who has proofread much of the material and made many helpful suggestions during the course of preparation of the manuscript.

L. S. Birks

Washington, D. C.
October, 1959

FOREWORD TO FIRST EDITION

Although X-ray spectrochemical analysis has attained widespread popularity only in the past decade, it is by no means a new field of interest. Study of the X-ray spectral lines and their relation to the chemical elements began shortly after the discovery of X-rays by Röntgen in 1895. In 1911, Barkla and his co-workers showed that characteristic radiation was emitted from an element when it was stimulated by X-radiation of "slightly shorter wavelength" than that which was emitted. Kay, Whiddington, and others soon found that the same characteristic radiation was emitted when the element was made the target in an X-ray tube and bombarded with electrons of sufficiently high energy. In 1913, Moseley began his classical experiments on the relation of wavelength to atomic number and established the simple relation that bears his name: "The frequency of the characteristic lines is proportional to the square of the atomic number." As the theory of atomic structure developed, it was apparent that Moseley's Law was part of the general picture of emission during change in the quantum energy state of an atom.

During the 1920's, von Hevesy and other workers became interested in the use of characteristic X-rays for chemical analysis. Solids or powders to be analyzed were placed on the targets of X-ray tubes and their spectra photographed in X-ray spectrographs using flat calcite crystals to disperse the spectra. A milestone in X-ray spectroscopy was reached in 1923 when Coster and von Hevesy established conclusive proof of the existence of element 72 (Hf) from X-ray spectra of Norwegian zircon. By the 1930's, concentrations of 10^{-4} or 10^{-5} could be measured using suitable standards, and in his book in 1932, von Hevesy stated that he looked forward to the day when concentrations of 10^{-6} could be measured.

The full development of the X-ray method had to wait, however, for technological advances. It was not until the mid 1940's that high-powered, stable, sealed-off X-ray tubes, large single crystals of synthetic rock salt, and sensitive Geiger counters were all brought together to make possible rapid and reproducible measurement of X-ray spectra excited by fluorescence. Great strides have been made in improvements in components and techniques during the past decade with the result that X-ray spectrochemical analysis is now a well established analytical tool that supplements and extends the range of other tools for chemical analysis.

It is the purpose of this volume to try to bring the subject of X-ray spectrochemical analysis up to date and to point out some of its future possibilities.

L. S. B.

CONTENTS

CHAPTER 1

SIMPLIFIED FUNDAMENTALS

Each chemical element emits characteristic X-ray lines when excited by high-energy radiation. By measuring the wavelength and intensity of these characteristic lines, one can tell what elements and how much of each are present in a sample. Fluorescent X-ray spectrometry is a particularly simple technique for such analysis. The following short description will introduce the newcomer to the subject and point out the processes to be considered in detail in succeeding chapters.

1.1. Relation of X-Ray Wavelength to Atomic Number

Practical use and attractiveness of X-ray spectrometry depends to a considerable extent on (*a*) the small number of lines in the characteristic X-ray spectrum of each chemical element, and (*b*) the very simple relationship between wavelength, λ, and atomic number, Z. The general relation is [1,2]

$$\lambda \propto 1/Z^2 \tag{1-1}$$

where the proportionality constant depends on the series, i.e., K, L, etc., which will be explained in detail in Chapter 2. For instance, the equation for the K_α lines is

$$\lambda \approx 1300/Z^2 \tag{1-2}$$

Figure 1-1 shows a log–log plot of λ versus Z for the useful K, L, and M lines for elements from B-5 to U-92. Details of the K series of V and the L series of W are shown as the inserts of Fig. 1-1 and illustrate the relative intensities and wavelengths.

There are some near overlaps of L or M lines of higher atomic numbers with K lines of lower atomic numbers but present spectrometers will usually resolve them or they will be distinguishable by the number and intensity of lines or by previous knowledge of which elements are likely to be present.

1.2. Geometry of Fluorescent X-Ray Spectrometers

The simplest arrangement of components for fluorescent X-ray spectroscopy is shown schematically in Fig. 1-2. The main parts are A, the primary X-ray tube; B, the specimen; C, the collimator; D, the

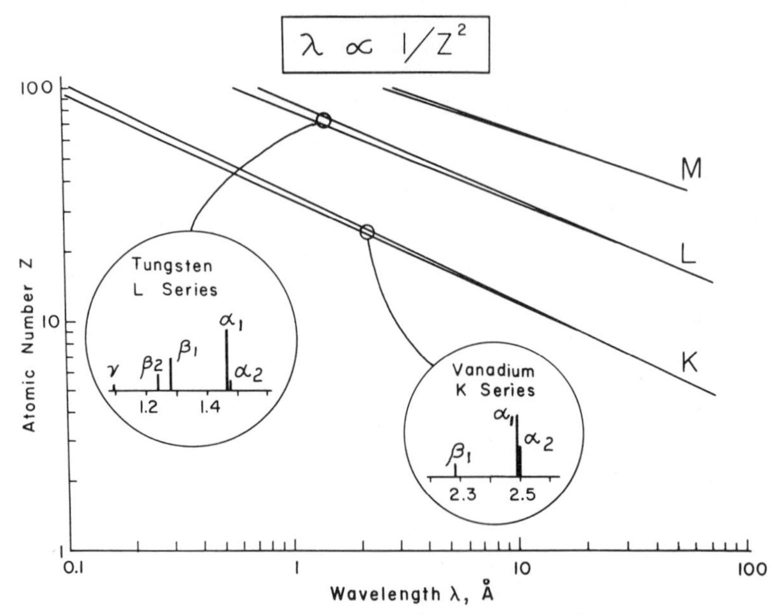

Fig. 1-1. Atomic number vs. wavelength.

analyzing crystal; *E*, the detector. Primary X-rays from the X-ray tube strike the specimen and generate the characteristic X-rays of the specimen elements. The specimen may be a single element such as a sheet of copper, an alloy such as steel, a mixture of powders such as paint pigments, or a liquid containing certain elements in solution. The characteristic X-rays are emitted in all directions and the first step in analysis is to limit them to a parallel bundle using the collimator.

The collimator is usually an array of parallel blades with spacings as small as 0.005 in. or as large as 0.050 in. Angular divergence of the emerging beam varies between 0.07 and 0.7° depending on the spacing. As will be shown in Chapter 4, the divergence allowed by the collimator is usually the limiting factor in the resolution of the spectrometer.

After passing through the collimator, the nearly parallel bundle of polychromatic radiation strikes the analyzing crystal. For each setting of the crystal, only one wavelength will be diffracted according to the Bragg equation.[2,3]

$$n\lambda = 2d \sin \theta \tag{1-3}$$

where n is the order of diffraction (we will be concerned primarily with the first order: $n = 1$), λ is the wavelength in angstroms, d is the inter-

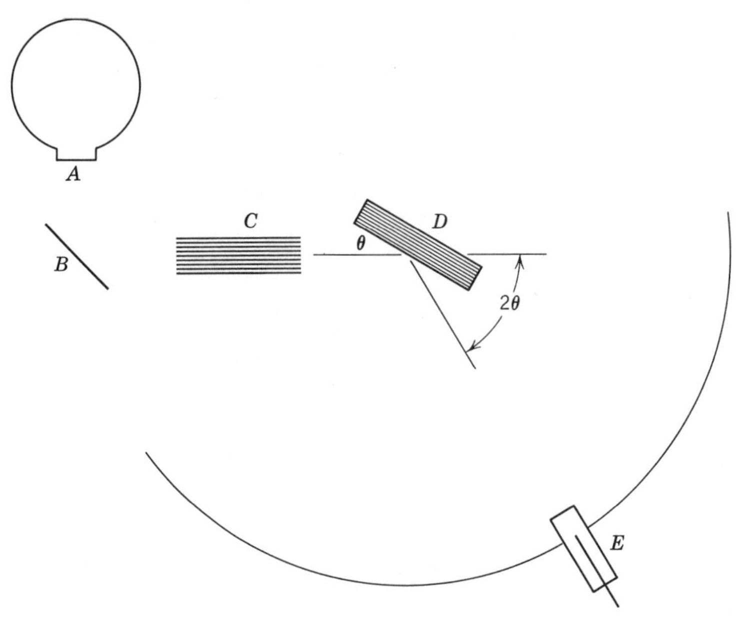

Fig. 1-2. Flat crystal spectrometer geometry: (*A*) X-ray tube, (*B*) specimen, (*C*) collimator, (*D*) analyzing crystal, (*E*) detector.

planar spacing of the crystal in angstroms, and θ is the angle between the incident radiation and the crystal surface as shown. The diffracted radiation emerges at an angle of 2θ with respect to the incident beam and is measured by the detector. To measure the whole spectrum from the specimen, the crystal is turned through the range $\theta = 0°$ to $\theta = 90°$, and the detector is turned at twice the speed of the crystal so it will always be in position to intercept the diffracted radiation. The wavelengths of the measured X-ray lines determine the elements present in the specimen, and the intensity of each line is related to the percentage composition of that element.

1.3. Plan of Succeeding Chapters

This book is intended as an introductory book for analytical chemists who will use fluorescent X-ray spectrometry as a standard tool for quantitative analysis and for metallurgists, mineralogists, biologists, etc. who need a basic understanding of the X-ray method in order to take advantage of its capabilities for their particular specimens. Therefore, Chapter 2 on the principles of X-rays deals only with those simplified physical concepts of generation, diffraction, and absorption which are a useful background to practical application of the method. Chapters 3, 4, 5, and 6 cover the mechanics of obtaining and measuring X-ray spectra. Chapters 7 and 8 show how the data are interpreted and quantitative analysis achieved. Chapter 9 is devoted to practical specimen preparation and applications. Chapter 10 is a condensed coverage of the specialized technique of electron probe microanalysis. Appendices 1, 2, 3, and 4 contain tables and other useful information.

References

1. H. G. J. Moseley, *Phil. Mag.*, **26**, 1031 (1913).
2. E. F. Kaelble, Ed., *Handbook of X-Rays*, McGraw-Hill, New York, 1967.
3. W. L. Bragg, Ed., *The Crystalline State*, Bell, London, 1933.

CHAPTER 2

PRINCIPLES OF X-RAY GENERATION, DIFFRACTION, AND ABSORPTION

2.1. What are X-Rays?

X-rays are like light, heat, or radio waves, that is, they are electromagnetic radiation but of very short wavelength. Short wavelength means high energy. One simple rule, $E = h\nu$, applies to all electromagnetic radiation; that is, energy, E, is equal to Planck's constant, h, times the frequency, ν. Of course ν is related to wavelength, λ, by the velocity of light, C; $\nu = C/\lambda$. This rule is used constantly by X-ray analysts in determining what X-ray tube voltage is necessary to excite certain elements, in deciding whether or not radiation from one element will excite radiation of another element, in relating absorption edges and emission lines, etc. It is usually expressed in practical working units with the constant adjusted accordingly.

$$\lambda = 12400/V \tag{2-1}$$

λ is in angstroms and V is in electron volts. For an X-ray photon of 1 Å wavelength the energy is thus 12400 eV; for a photon of green light $\lambda = 5000$ Å and $V \approx 2.5$ eV; for a one-meter radio wave $\lambda = 10^{10}$ Å and $V \approx 1.2 \times 10^{-6}$ eV.

Why are X-rays so useful in studying the chemical and physical properties of materials? There are two main reasons:

(*a*) The X-ray energies represent the energy levels of inner electrons in atoms and, as was stated in Chapter 1, are uniquely related to atomic number. Therefore they distinguish one element from another and this is what we use for fluorescent X-ray analysis.

(*b*) The wavelengths of X-rays are about the same as the spacings between atoms in solids or liquids. Thus by measuring how X-rays are scattered or diffracted we can tell how atoms are arranged in alloys, molecules, and especially in crystalline compounds. This is X-ray diffraction.

2.2. Origin of Characteristic X-Rays

A characteristic X-ray photon is generated by a two-step process. The first step is illustrated in Fig. 2-1 where a high-energy quantum such

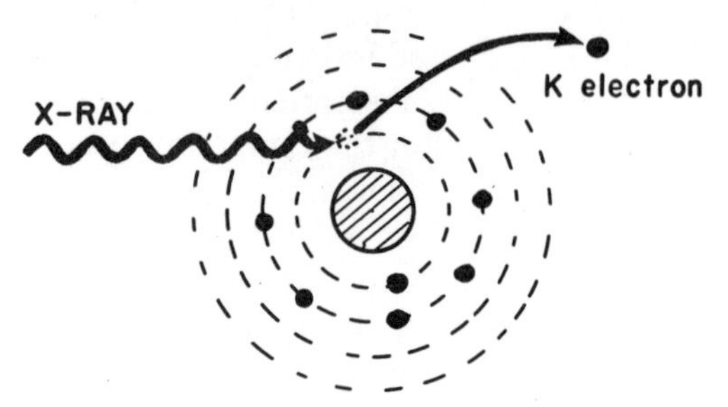

Fig. 2-1. Schematic of the removal of a K electron from an iron atom by a primary X-ray photon.

as an electron, an X-ray photon, or a proton strikes an atom and knocks out an inner-shell electron. The second step is readjustment in the atom almost immediately (10^{-12} to 10^{-14} sec) by filling the inner-shell vacancy with one of the outer electrons and simultaneous emission of an X-ray photon. The first step uses up the energy of the incident quantum, and in the second step energy is emitted as the characteristic X-ray photon. The incident quantum may have any energy greater than the binding energy of the inner-shell electron with the excess energy carried away as kinetic energy of the electron being removed. But replacement of the inner-shell electron by one of the outer-shell electrons corresponds exactly to the difference in energy between the two levels; hence the emitted photon has the characteristic energy of the difference in levels.

The simplest representation of energies in atoms, and a useful means of explaining the several lines in a series, is the energy level diagram. A partial diagram[1] is shown in Fig. 2-2. It is not to scale but only a pictorial representation. Allowed transitions are given by the selection rules of quantum mechanics,[2] but, simply stated, they say that the second quantum number must change by ± 1. The second quantum number, s, p, d, f, etc., designates the level within the main shell which is given by the first quantum number, K, L, M, etc. From Fig. 2-2 it is seen that when a $K(1s)$ electron is removed it may be replaced by a p electron from the L or M shell. Replacement by a $2p$ electron from the L_{III} shell leads to emission of a K_{α_1} characteristic X-ray photon; likewise, replacement by a $2p$ electron from the L_{II} shell leads to emission of a K_{α_2} photon. Replacement by one of the $3p$ electrons leads to one of the K_β lines. The difference in energy between $3p$ and $1s$ being greater than that between $2p$ and $1s$ means a greater photon energy for the K_β line than for the K_α line, i.e., a shorter wavelength according to Eq. 2-1.

Removal of the K electron corresponds to the difference in energy between the $1s$ level and the ground level or zero. This energy difference is called the characteristic K absorption energy, V_K, and is always greater than the energy of any of the characteristic emission lines.

If the original electron removed was from the L shell, the replacement would have to be from the M or N shell, etc. Since there are three levels in the L shell there are three nearly equal L absorption energies, $V_{L_I}, V_{L_{II}}, V_{L_{III}}$. Replacement of the $2s$ electron must be by a p-type electron but replacement of a $2p$ electron may be by either an s or d type electron according to the selection rules. For the M series a similar situation prevails but it is more complex.

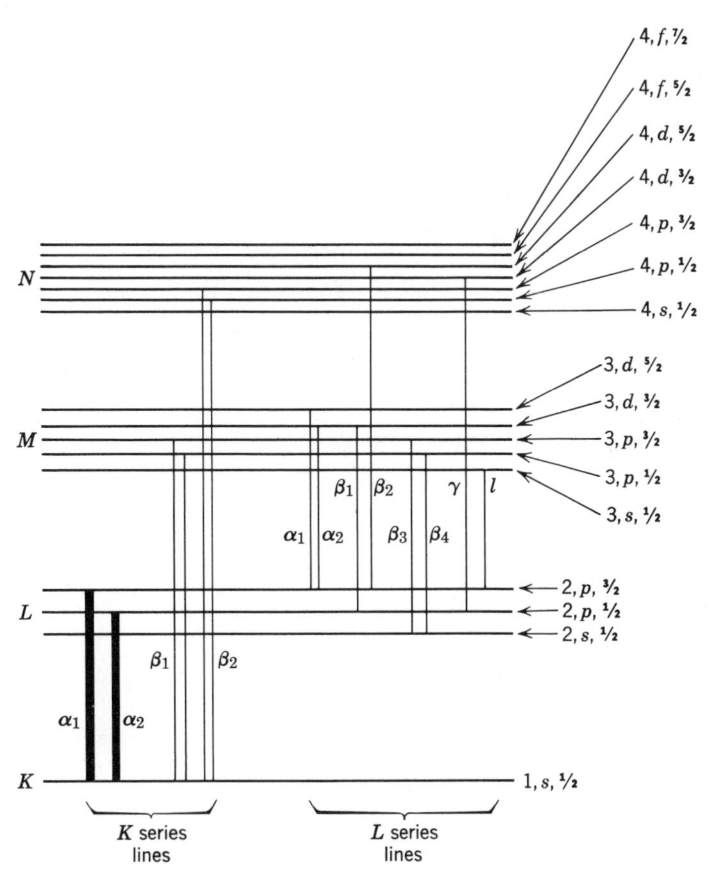

Fig. 2-2. A partial energy level diagram showing the transitions leading to the K and L series lines.

The intensities of the various lines in a series depend on the probability that replacement will occur from a particular outer level. Certain generalities can be stated such as K_{α_2} is half as intense as K_{α_1}, and K_β is about 1/5 as intense as K_{α_1}, but they are not exact. In the L series the L_{α_1} is stronger than the L_{β_1} for most elements but vice versa for some elements. Tables of relative intensities are available.[3] The naming of characteristic lines occurred before about 1925 before quantum mechanics was well established; it was done initially on the basis of intensity. Thus the names of the individual L_β lines, for instance, do not always appear logical from a quantum mechanics view point but are firmly established in history.

2.3. Origin of the Continuous X-Ray Spectrum

If the incident quantum is an electron it does not necessarily knock out an inner-shell electron but instead may lose all or any part of its energy by interaction with the nucleus of the atom. In any case, the X-ray photon emitted in such a process has the same energy as that lost by the passing electron. In an X-ray tube the variable loss of energy by electrons leads to a continuous spectrum[4] of X-ray photon energies originally called Bremsstrahlung or "white" radiation. The minimum wavelength in the continuous spectrum corresponds to the maximum electron energy available and is independent of the atomic number of

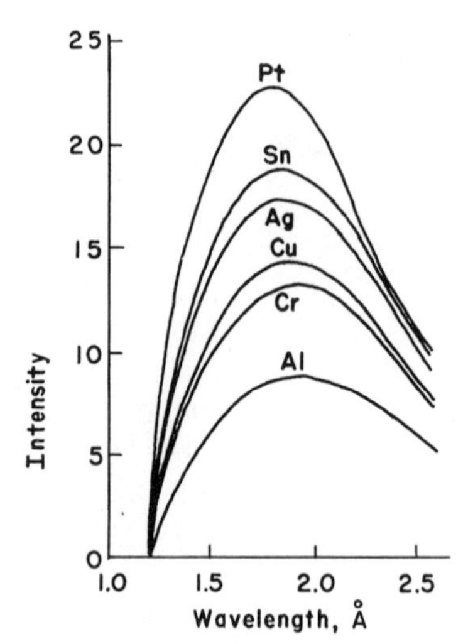

Fig. 2-3. Continuous X-ray spectra from several X-ray targets all at 10 keV. (After Kulenkampff, Ref. 6)

the target element.[5] The relative spectral distribution in the continuum is also similar from element to element but the absolute intensity is greater for higher atomic numbers.[6] Figure 2-3 shows a portion of the continuum for six different elements. Equation 2-2 gives an approximate relation[4] which is most conveniently expressed in terms of energy V, rather than wavelength, λ.

$$I_V = AZ(V_p - V) + BZ^2 \qquad (2\text{-}2)$$

where I_V is the intensity at energy V, A and B are empirical constants ($A \gg B$), V_p is the maximum (peak) voltage on the tube, and Z is atomic number.

It should be noted that the continuous spectrum is observed only for electron excitation. For proton excitation the continuum is weaker by about the ratio of the square of proton mass to electron mass, i.e., about $(1800)^2$ times less intense and not observable practically. For incident X-ray photons as in the case of fluorescent X-ray spectrometry, there is no variable energy loss and therefore no continuum because X-ray photons lose all or none of their energy.* The background observed in fluorescence X-ray spectrometry is merely primary radiation scattered from the specimen or, to a lesser extent, characteristic X-rays emitted by the collimator or crystal.

2.4. X-Ray Diffraction

Only the rudiments of X-ray diffraction necessary for understanding the crystal analyzers in X-ray spectrometers will be described.

Crystals are regular arrays of atoms and act like three-dimensional gratings to diffract X-rays.[7] Bragg's equation relates wavelength, λ, to diffraction angle, θ, and spacing between planes of atoms, d. Figure 2-4 shows one set of planes in a crystal and the diffraction situation schematically. Diffraction occurs when there are an integral number of wavelengths, $n\lambda$, between radiation scattered by adjacent planes of atoms. In Fig. 2-4 the additional path length necessary for radiation to reach the second plane is given by l where $l/d = \sin\theta$. The same additional path length occurs for the scattered radiation emerging from the second plane. Thus, the total additional path length is $2l = 2d \sin\theta$. For diffraction to occur, $2l$ must equal $n\lambda$; hence we have

$$2l = n\lambda = 2d \sin\theta \qquad (2\text{-}3)$$

which is the familiar Bragg law.[7]

* Compton (incoherent) scattering does correspond to a slight degrading of photon energy but does not result in emission.

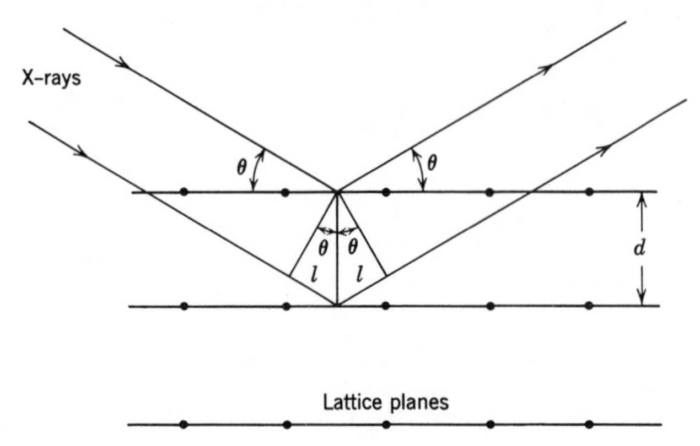

Fig. 2-4. Diffraction from lattice planes according to the Bragg law.

X-ray diffraction by a crystal differs from visible-light diffraction by a plane grating in that only one X-ray wavelength, λ (and higher orders of $\lambda/2$, $\lambda/3$, etc.) out of a polychromatic incident beam will be diffracted at a time. First-order diffraction where $n = 1$ is the most intense but second and third and higher orders do occur in X-ray spectrometers and can cause interference. For instance, the Hf L_α line at 1.566 Å will be diffracted by a LiF crystal in the first order at 22.9° θ, while Zr K_α at 0.78 Å will be diffracted in the second order from the same crystal at 23.0° θ. However, the Zr K_α has about twice the energy of Hf L_α and its interference can be eliminated electronically by energy discrimination as described in Chapter 6.

Each crystal has many different sets of planes with different spacings but we will be concerned primarily with the planes parallel to the surface of the crystal analyzer. Also, different crystals have different interplanar spacings and therefore diffract at different angles. Details of the characteristics and choice of crystals is given in Chapter 4.

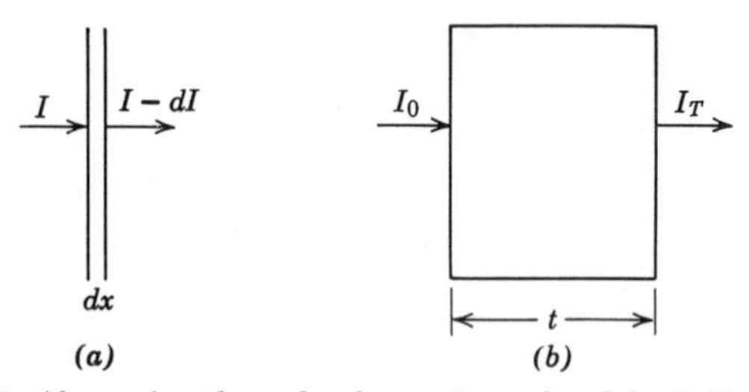

Fig. 2-5. Absorption through a layer, dx, and a slab of thickness t.

2.5. X-Ray Absorption and Scattering

X-ray absorption, like diffraction, is a subject in itself and only the rudiments will be given.

X-rays obey the same absorption law as other electromagnetic radiation. Figure 2-5 shows the situation. In Fig. 2-5*a*, the incremental loss of intensity, dI, in passing through an incremental layer, dx, is proportional to the intensity, I.

$$dI \propto I dx$$

The constant of proportionality is usually written as μ and called the linear absorption coefficient. Rewriting and integrating over the limits of Fig. 2-5*b* we have

$$dI/I = \mu dx$$

$$\ln I \Big|_{I_0}^{I_T} = \mu x \Big|_0^t$$

$$I_T/I_0 = \exp(-\mu t) \tag{2-4a}$$

$$= \exp[-(\mu/\rho)\rho t] \tag{2-4b}$$

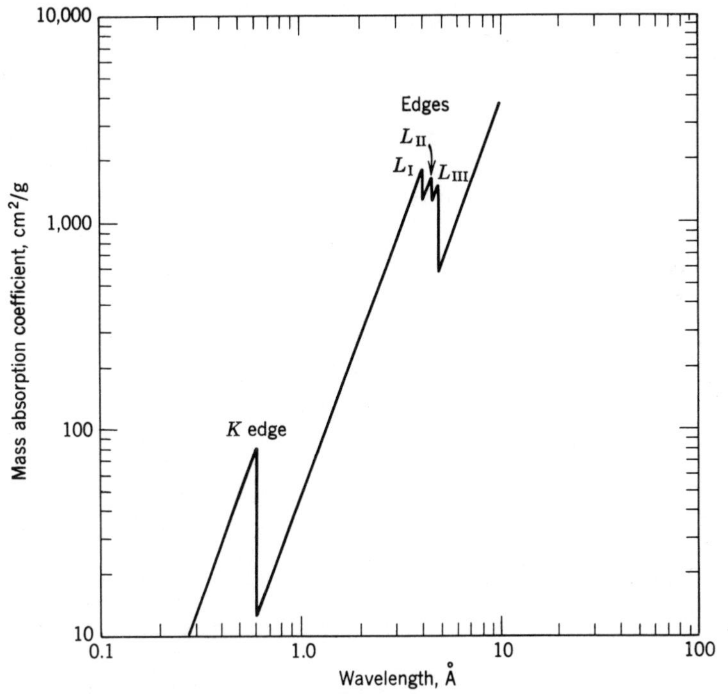

Fig. 2-6. Mass absorption coefficient vs. wavelength for Mo.

The mass absorption coefficients, μ/ρ, where ρ is density, are much more useful than the linear coefficients μ and customarily are the ones tabulated. Nowadays some writers (including this one) use just the term μ for mass absorption especially when it has subscripts to describe absorption of a particular element for a particular wavelength.

For a given element the mass absorption coefficient varies inversely with wavelength approximately as λ^3 except at the characteristic absorption edges where there are sharp jumps. Figure 2-6 shows the curve of μ/ρ for Mo. The absorption edges shift monotonically in wavelength with atomic number. Mass absorption coefficients are additive linearly. That is, if the slab in Fig. 2-5 consists of a matrix, M, of several elements with weight fractions W_1, W_2, W_3, etc. then the total mass absorption coefficient, $\mu_{M\lambda}$, of the matrix for wavelength λ is expressible as

$$\mu_{M\lambda} = \mu_{1\lambda}W_1 + \mu_{2\lambda}W_2 + \mu_{3\lambda}W_3 + \ldots \tag{2-5}$$

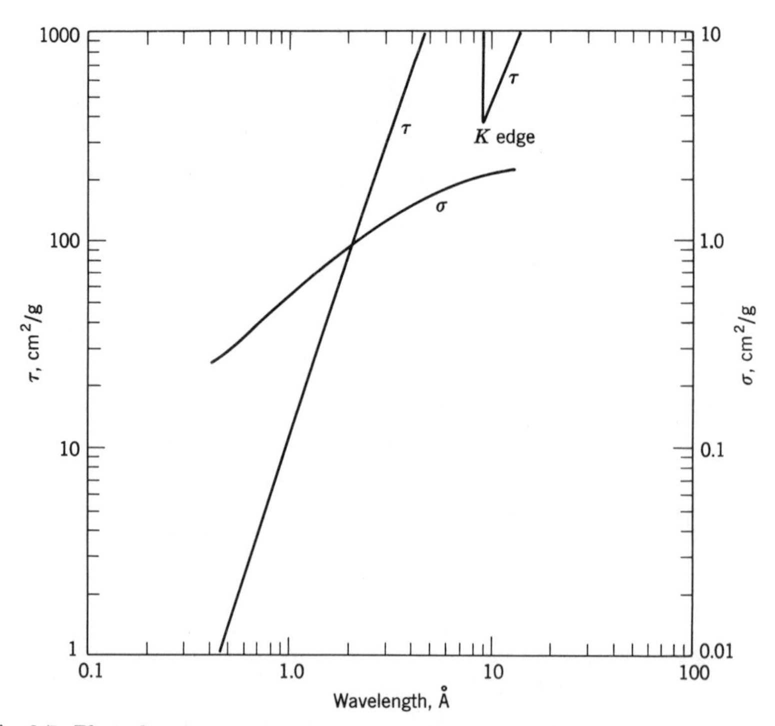

Fig. 2-7. Photoelectric component, τ, and scattering component, σ, for the mass absorption coefficient of Mg.

The mass absorption coefficient is really made up of two components, the photoelectric or true absorption, τ, and the scattering, σ. The photoelectric part is the predominant part (over 95%) for most X-ray fluorescent work as shown in Fig. 2-7 for Mg. For very low atomic number elements such as C, N, and O, scattering, σ, becomes predominant.[8] Scattering is important in calculating the intensity generated by X-ray transport in low Z materials and in other problems in X-ray analysis.

References

1. H. A. Liebhafsky, H. G. Pfeiffer, E. H. Winslow, and P. D. Zemany, *X-Ray Absorption and Emission in Analytical Chemistry*, Wiley, New York, 1960, p. 32.
2. A. H. Compton and S. K. Allison, *X-Rays in Theory and Experiment*, Van Nostrand, New York, 1935, p. 629.
3. M. Siegbahn, *Spektroskopie der Rontgenstrahlen*, Springer, Berlin, 1931.
4. A. H. Compton and S. K. Allison, *X-Rays in Theory and Experiment*, Van Nostrand, New York, 1935, p. 38.
5. W. Duane and F. L. Hunt, *Phys, Rev.*, **6**, 166 (1915).
6. H. Kulenkampff, *Ann. Physik*, **69**, 548 (1922).
7. W. L. Bragg, Ed., *The Crystalline State*, Bell, London, 1933.
8. W. H. McMaster, N. Kerr Del Grande, J. H. Mallett, N. E. Scofield, R. Cahill, and J. H. Hubbell, Lawrence Radiation Laboratory of the University of California, Report UCRL 50174, 1967.

CHAPTER 3

EXCITATION FOR X-RAY ANALYSIS

In this chapter we are concerned with excitation of characteristic radiation from the elements in a specimen. Characteristics of the exciting sources are discussed only in relation to this primary concern.

3.1. Spectral Distribution of Primary Radiation in X-Ray Tubes

The usual source of excitation in fluorescent X-ray analysis is the primary spectrum from an X-ray tube. The spectral distribution in the primary spectrum is important because it determines how well the characteristic radiation will be excited in the specimen. The only effective part of the primary spectrum is that part which is of shorter wavelength than the absorption edge, λ_{abs} of the element to be excited in the specimen. Target material and voltage determine the intensity in the effective part of the primary spectrum.

Spectral distribution for W and Cr target fluorescent X-ray tubes are shown in Fig. 3-1. (The peak heights of the characteristic lines are indicated.) These distributions represent the radiation outside the tube window and available to excite the specimen. Appendix 1 tabulates the spectral distributions for W, Cu, Cr, and Mo target fluorescent tubes and the characteristic line intensities are included.

Two features are noticeable about the spectra:

(*a*) The characteristic lines contribute a large part of the total radiation.

(*b*) The jump in the continuum at the characteristic absorption edge for the target element means more primary intensity available at longer wavelengths than had been commonly realized prior to 1967.[1] This jump is caused by reduced absorption of the radiation on the long wavelength side of the edge as it emerges from below the target surface. It explains part of the observed effectiveness of Cr target tubes for exciting low atomic number elements. Adjustment of the spectra shown can easily be made for tubes with different windows by correcting for the different absorption at each wavelength. There is no very accurate way to adjust for other voltages although Eq. 2-2 of the previous chapter can be used to make rough estimates.

Figure 3-2 shows the tungsten spectrum at 50 kV full-wave rectified potential. Compared to Fig. 3-1 there is less intensity in the short wavelength portion of the continuum. This may or may not be important depending on the elements to be excited in the specimen. The reason for the difference is illustrated in Fig. 3-3. V_p is the peak voltage and V_{abs} is the minimum energy necessary to excite the K lines of some

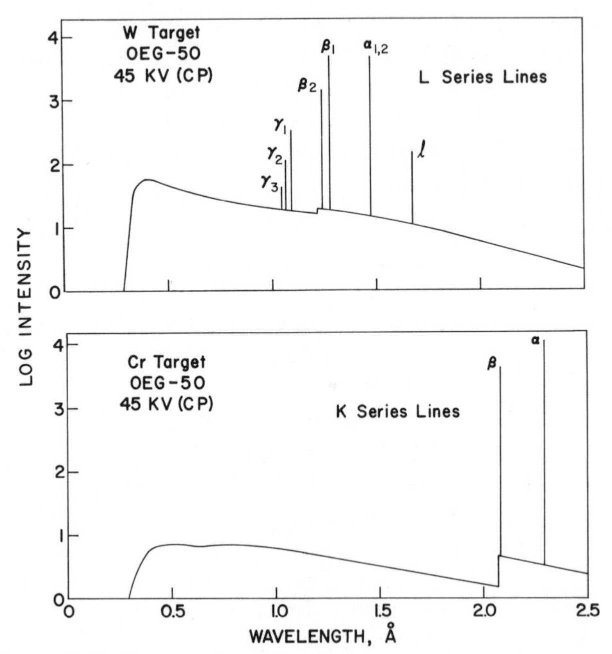

Fig. 3-1. Spectral distributions from W and Cr target X-ray tubes operated at constant potential (CP).

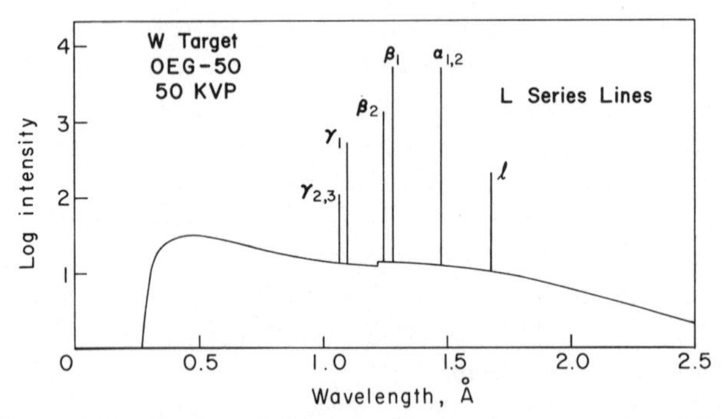

Fig. 3-2. Spectral distribution of W target X-ray tube operated at full-wave rectified potential.

hypothetical element in the specimen. Only that part of the cycle above V_{abs} will produce primary radiation which can excite the specimen. The more closely V_{abs} approaches V_p, the less effective the tube will be. Figure 3-4 shows the relative increase in emission for constant potential operation. I_{CP}/I_{rect} is the relative increase in intensity and V_{abs}/V is the ratio of the values shown in Fig. 3-3. The calculated curve is in good agreement with experimental measurements by Spielberg and Mack.[2]

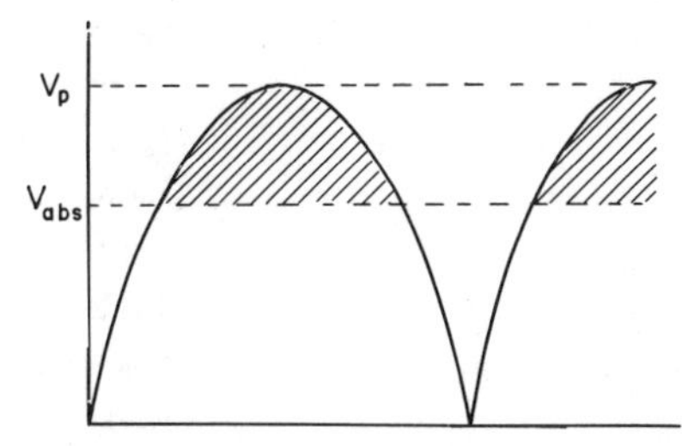

Fig. 3-3. Idealized form of full-wave rectified potential applied to an X-ray tube. The only effective portion of the cycle is shown by the shaded portion where the tube potential is above V_{abs}.

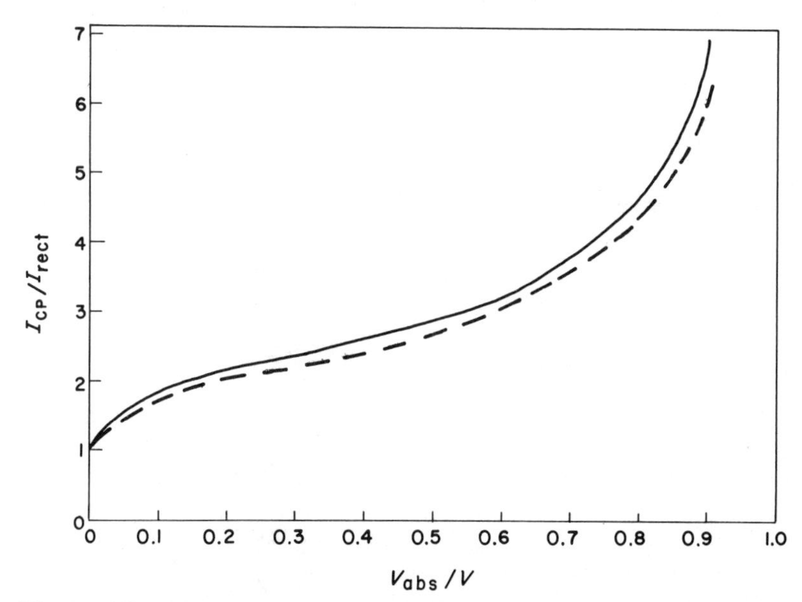

Fig. 3-4. The relative increase in expected characteristic X-ray intensity for constant potential operation compared to full-wave rectified operation (observed values from Spielberg and Mack, Ref. 2). —— calculated. - - - observed.

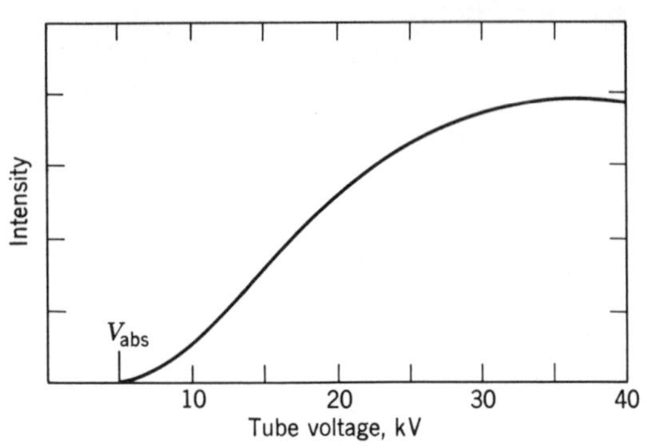

Fig. 3-5. Approximate variation of characteristic line intensity with tube potential (CP).

A few simple statements may be made about primary spectral intensity and tube voltage. For the characteristic lines in the target spectrum the intensity for constant-potential operation varies approximately as shown in Fig. 3-5. For $V < V_{abs}$ there are no characteristic lines generated. For $V < 2$ or $3\ V_{abs}$ the increase in intensity is about as $(V - V_{abs})^2$. For $V > 3V_{abs}$ the increase in intensity becomes more nearly linear up to some high value of V when the generation is so deep in the target that less of the radiation can emerge and the intensity begins to decrease. In the case of the continuum, the intensity variation with voltage at some wavelength λ_i is approximately linear with overvoltage, $V_p - V_i$, where V_p is the peak voltage on the tube and V_i is the voltage corresponding to λ_i (see Eq. 2-2).

3.2. Intensity of Characteristic Lines from the Specimen as Related to Primary Spectrum

As was stated in Sec. 3.1, only that part of the primary spectrum is effective which is of wavelength shorter than λ_{abs} of the specimen element to be excited. Figure 3-6 illustrates the excitation of a Cu specimen by a W target X-ray tube. The left-hand series of pictures shows the primary spectrum for a series of tube voltages and the right-hand series shows the corresponding Cu intensity from the specimen. Note that intensities are only relative; one should not attempt to relate absolute intensities from left-hand to right-hand graphs. At 6 kV tube potential, none of the primary spectrum will excite Cu in the specimen. As the tube voltage is increased to 12, 25, and 50 kV, more and more of the primary spectrum is effective and the Cu fluorescence intensity rises rapidly. Generally speaking, the tube should be operated at 3 to 10

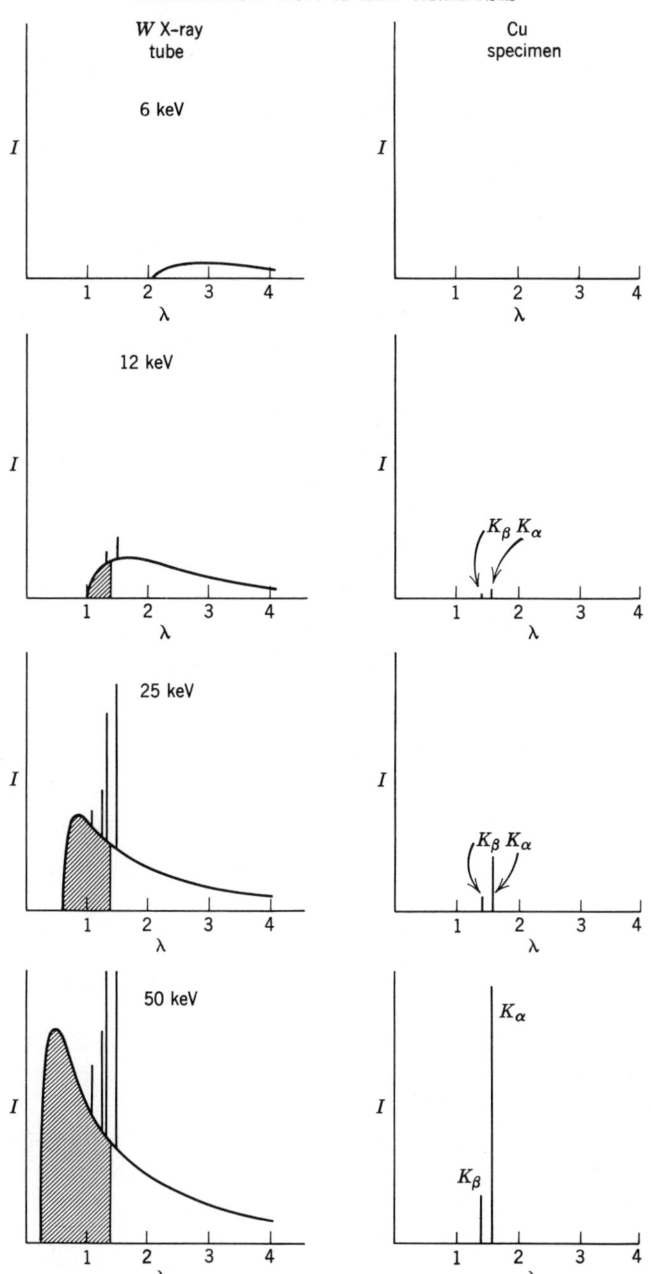

Fig. 3-6. Effect of tube voltage on the primary spectral distribution in a W target tube and the corresponding characteristic intensity excited in a Cu specimen. Note: the intensity scales are not the same for primary and secondary intensities.

times the voltage corresponding to λ_{abs} of the specimen element. This is not always possible because the voltage of most tubes is limited to about 50 kV. Thus, less than a factor of $2V_{abs}$ times is reached for Sn.

3.3. Choice of Tube and Target

The choice of X-ray tube for optimum excitation of a particular type of specimen depends on several factors which must be judged relative to each other.

(*a*) Only a few elements have suitable thermal and mechanical properties for X-ray tube targets. W is the most common target material because it has a very high melting point and reasonable thermal conductivity for cooling, and is of high atomic number so that it gives an intense spectrum (see Eq. 2-2). Pt, Mo, Ag, Cu, Fe, and Cr are also suitable but more limited than W in the wattage at which they may be operated. One might ask why W is not used exclusively since it can withstand greater wattage and gives a stronger spectrum than lower atomic number elements. One reason is the direct interference of the W L lines with the characteristic K or L lines of various elements in the specimen. Second is the deliberate use of the characteristic lines of other targets to enhance the excitation of particular elements in the specimen. For instance, Ag is ideal for exciting Mo, Cu for exciting Fe, Ag L lines for exciting Cl and S; Cr for exciting Ti, etc. Therefore, if repeated analyses are to be performed on the same system of elements, it is wise to consider which tube target will be most effective for excitation.

(*b*) Tube window thickness is important especially for exciting low atomic number elements. Sulfur, for instance, is excited about 250 times more effectively by a 0.010 in. Be window tube than by the standard 0.040 in. Be window tube because of the absorption of the effective portion of the primary spectrum by the Be. However, the thin-window tubes are more fragile, more expensive, and usually do not last as long, so if the specimen elements of interest are above Cr-24 a regular window tube is probably a better choice.

(*c*) Some tubes are constructed so that the focal spot is off center, closer to the window. This increases available intensity because the solid angle of primary radiation intercepted by the specimen varies inversely as the square of the distance between specimen and focal spot. A shift of 2 cm in the focal spot would increase intensity by 2.8 times for a specimen located 2 cm from the window. The off-center tubes cannot be operated at as high a wattage as normal tubes because the window area becomes hotter. In addition, there is likely to be greater deposition of tungsten from the filament on the window; this will reduce available intensity more rapidly with age.

(*d*) Some tubes have multiple targets such as W and Cr so that the most suitable one can be used for different types of specimens. The cost of one multiple-target tube is $2150 compared to $1000 for a W target and $1300 for a Cr target tube (1967 prices). If the multiple-target tube burns out or breaks, it is like losing two tubes and may mean stopping operation.

From the considerations in (*a*) through (*d*) above, it is easy to understand why no one tube is optimum for all analytical work. At the present time perhaps the best minimum complement of tubes would be a standard W target tube and a thin-window Cr target tube or a combination W and Cr target tube because these give reasonable coverage for most situations.

3.4. Other Methods of Excitation

Fluorescent excitation with an X-ray tube became, and still is, the most common method of X-ray spectrochemical analysis because it offers a good compromise on cost, easy sample interchange, and high intensity. However, excitation by radioactive sources or by direct electron or proton bombardment offer valuable features which make them better choices for some analyses.

Table 6-2 of Chapter 6 lists a variety of radioactive sources and their properties.[3] There are two principal types, those that emit γ rays such as ^{241}Am, ^{109}Cd, ^{57}Co, and those that emit electrons (β particles) which are converted to X-rays within the source such as ^{3}H + Ti or ^{3}H + Zr. The β emitters would be ideal for more situations if it were feasible to introduce hydrogen into any chosen element, but it is not feasible. Advantages of the radioactive sources are their small size and low cost compared to the initial cost of X-ray tubes and power supplies. Disadvantages include low intensity (see Table 6-2) and short lives for some of the isotopes. For instance, the intensity is so low compared to X-ray tubes that radioactive sources cannot be used with dispersive X-ray spectrometers but only with energy dispersion (often called nondispersive) equipment discussed in Chapter 6. Specific examples of applications are given in Chapter 9 along with all the other applications.

Direct electron excitation offers a 20,000 increase in intensity over fluorescent excitation[4] for the same total power but requires that the specimen be introduced into the vacuum system of the electron source.

Such a procedure is not especially difficult with present high-speed vacuum systems and, of course, is the method for all electron probe microanalysis (see Chapter 10). The background interference from the continuous spectrum limits the detectability to about 20–100 ppm for middle atomic number elements[5] as compared to 1–10 ppm for fluorescent excitation.[6] For low atomic number elements where sufficient intensity is always an important problem, direct electron excitation offers some advantage.

Proton excitation has been investigated[7] and found feasible in some instances. There is no measurable continuum for proton excitation which is an advantage over electron excitation. Generally speaking, proton sources are quite expensive compared to electron sources or X-ray tubes. Originally it was thought that protons would be more effective than electrons for exciting X-rays in thin foils because their penetration would be less. However, the proton energy is about 100

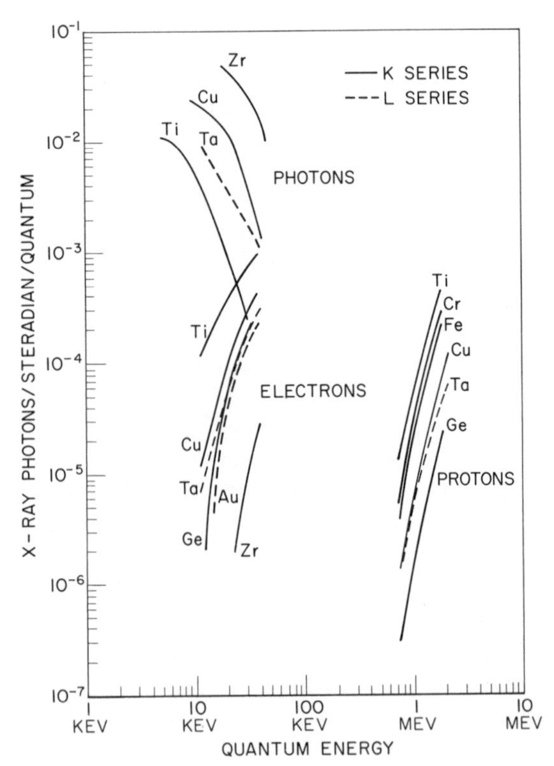

Fig. 3-7. Total X-ray yields for excitation by protons, electrons, and primary X-ray photons.

times higher than electron energy for the same X-ray yield (see Fig. 3-7) and the penetration of the high-energy protons is actually greater than that of the lower-energy electrons.

3.5. Miscellaneous Factors Affecting Observed Intensities from Specimens

Intensities of the characteristic lines from elements in specimens depend on several factors in addition to the primary spectral distribution discussed in Secs. 3.1 and 3.2.

(*a*) *Fluorescent Yield:* As was explained in Chapter 2, whenever an inner electron is removed from an atom it must be replaced quickly from one of the outer shells. The X-ray photon generated does not necessarily escape from the atom, however; instead it may knock out one of the outer electrons of the same atom and thus generate L or M photons. This process is called the Auger effect[8] and the chance of its occurrence depends on atomic number. K fluorescent yield, ω_K, is the fractional probability of K emission per K ionization; for instance, on the average for every 100 Cu atoms ionized in the K shell there will be 44 Cu K photons emitted giving an $\omega_K = 0.44$. Likewise the L fluorescent yield ω_L is a similar probability for emission from L shell ionized atoms. Appendix 3 shows the average K and L yields from experimental observations.[9] There are uncertainties as large as 5–15% in the ω_K values and probably 10–20% in the ω_L values. The lower values of fluorescent yield for the low atomic number elements is one of the limitations in their determination.

(*b*) *Total Yield:* Total yield depends on the probability for causing ionization as well as on the fluorescent yield. It is thus strongly dependent on the type and energy of the incident quanta. Customarily one speaks of the yield in terms of intensity emerging from the specimen and therefore affected by absorption within the specimen. Figure 3-7 showed measured yield in terms of photons per steradian per incident quantum for excitation by electrons, protons, and X-ray photons. Several interesting features of the curves should be noted: (*1*) For electron or proton excitation the yield goes up as the energy of the incident quantum goes up but for X-ray excitation the greatest yield is for incident quanta with energies just greater than the absorption edge of the element being excited. This says that an incident X-ray with just enough energy to knock out an inner-shell electron has the greatest likelihood of doing so. (*2*) The crossover of the electron and photon curves tells us something about when it is better to excite with electrons and when it is better to excite with primary X-rays; electron excitation of low atomic number elements is favored. (*3*) The displacement of the electron and proton curves tell us that protons must have energies about 100 times higher than electrons in order to generate the same intensity. A side observation in the experiments leading to Fig. 3-7 was that the higher-energy protons actually penetrated deeper into the specimen than did the lower-energy electrons for the same yield. This was known by nuclear physicists[10] but not realized by most X-ray analysts.

(*c*) *Source Intensity:* The relation between source intensity and specimen intensity may seem so obvious that a discussion is superfluous. However, the extreme variation in numbers of incident quanta available from electron sources, X-ray tubes, and radioactive isotopes makes comparisons useful. For electron sources such as the electron probe it is possible to measure thousands of counts per second in spectrometers with electron currents as low as 10^{-7} to 10^{-8} A. For X-ray fluorescent sources (X-ray tubes) it takes about 10^{-2} to 10^{-3} A to achieve similar counting rates. With radioactive isotopes, it is usually not feasible to use spectrometers at all because the source strengths are lower by several more orders of magnitude; for isotope sources energy dispersion (Chapter 6) is the usual method of operation.

(*d*) *Matrix Absorption:* The observed X-ray intensity from a particular element in a specimen depends strongly on the absorption coefficients of the other elements for the characteristic radiation being measured. This matrix effect must be considered in quantitative analysis and is discussed in detail in Chapter 8.

3.6. Summary on Excitation

Fluorescent excitation by primary radiation from an X-ray tube is the most common method of excitation for X-ray analysis but radioactive sources, electrons, and protons are also useful for specialized applications. Spectral distribution in the primary radiation depends on tube target and voltage and in turn controls excitation of characteristic lines in the specimen. Tungsten is the most common tube-target material but other targets are also used. Observed intensities from the specimen depend on fluorescent yield, total yield, intensity of the source, and matrix absorption.

References

1. J. V. Gilfrich and L. S. Birks, *Anal. Chem.*, **40**, 1080 (1968).
2. N. Spielberg and M. Mack, *Norelco Reporter*, **5**, 109 (1958).

3. J. R. Rhodes, *Analyst*, **91**, 683 (1966).
4. R. C. Ehlert and R. A. Mattson, *Advan. X-Ray Anal.*, **9**, 457 (1966).
5. L. S. Birks, *Electron Probe Microanalysis*, Wiley, New York, 1963.
6. W. J. Campbell, J. D. Brown, and J. W. Thatcher, *Anal. Chem. Review Issue*, **38**, 416R (1966).
7. L. S. Birks, R. E. Seebold, A. P. Batt, and J. S. Grosso, *J. Appl. Phys.*, **35**, 2578 (1964).
8. E. H. S. Burhop, *The Auger Effect*, Cambridge Univ. Press (1952).
9. R. W. Fink, R. C. Jopson, H. Mark, and C. D. Swift, *Rev. Mod. Phys.*, **38**, 513 (1966).
10. J. Lindhard and M. Schorff, *Kgl. Danske Videnskab. Selskab, Mat-Fys. Medd.*, **27**, 15 (1953); J. M. Kahn, D. L. Potter, and R. D. Worley, Univ. of California, Lawrence Rad. Lab. Report UCRL-7826 (1964).

CHAPTER 4

DISPERSION: SPECTROMETER GEOMETRY AND CRYSTAL PROPERTIES

In this chapter we are concerned with dispersion of the characteristic radiation from the specimen according to its wavelength so that the intensity from each element can be measured separately. This is done with an X-ray spectrometer and an analyzing crystal. Energy dispersion (often called nondispersive geometry) which eliminates the spectrometer and crystal and relies on the proportional response of the detector is discussed in Chapter 6.

4.1. Flat-Crystal Spectrometers

(*a*) Flat, reflection crystal optics is probably the most common of all the X-ray fluorescence techniques. It was illustrated in Fig. 1-1 of Chapter 1 but is shown in more detail in Fig. 4-1. The blade collimator in Fig. 4-1*a* allows maximum divergence, $2\Delta\theta$, of $\pm$ arc tan (s/l) in the θ direction and, $2\Delta\phi$ of $\pm$ arc tan (p/l) in the ϕ direction parallel to the crystal axis of rotation. This corresponds to the segment of the diffraction cone shown in Fig. 4-1*b*. Except at 2θ angles near 180° the divergence in the ϕ direction has little effect on resolution, but divergence in the θ direction along with the crystal rocking curve breadth (Sec. 4.4) completely determines the resolution of the spectrometer. Common values for the collimator dimensions are l = 4 in., p = ¾ in., and s = 0.005–0.050 in. The intensity passed by the collimator is the triangular peak in Fig. 4-1*c*. Breadth of the peak at half maximum, B_c, is just arc tan (s/l). Combined effects of collimator divergence and crystal rocking curve are discussed in Sec. 4.5 and examples of resolution are shown.

(*b*) Flat crystals may also be used in transmission geometry as shown in Fig. 4-2; the diffracting planes are perpendicular to the surface. This is generally far less satisfactory than reflection geometry because the crystal must be very thin (usually $<$0.010 in.) to prevent strong absorption of the radiation.* Transmission geometry does have an

* Anomalous absorption in perfect Si or Ge crystals would allow transmission through crystals as thick as a few millimeters but the total diffracted intensity from perfect crystals would be less than 10% of the intensity from mosaic crystals such as LiF.

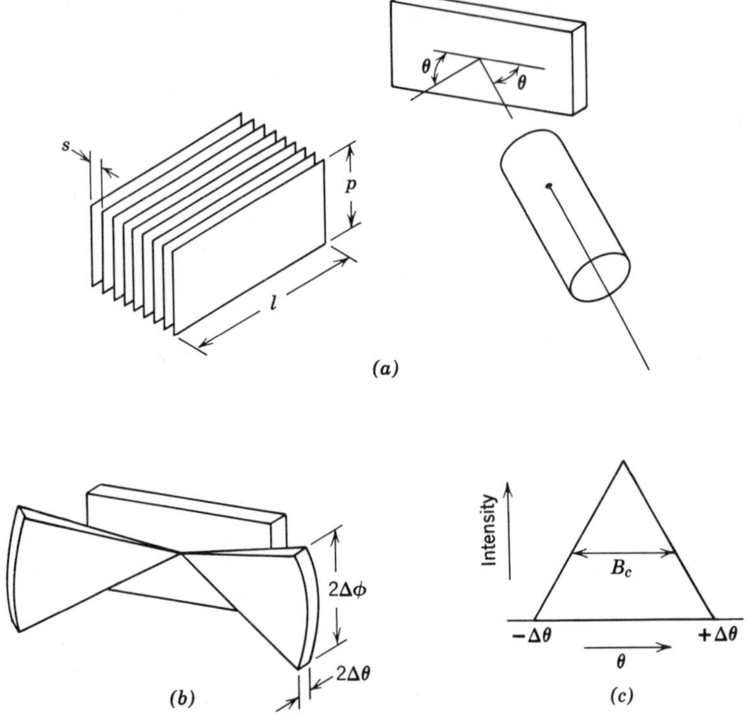

Fig. 4-1. Schematic of flat-crystal reflection optics. (*a*) Collimator, crystal, and detector. (*b*) A segment of the diffraction cone showing the spread $2\Delta\theta$ allowed by the blade spacing, *s*, and the spread, $2\Delta\phi$, allowed by blade height, *p*. (*c*) The intensity function obtained from a blade collimator.

advantage at very short wavelengths where the θ angle is small and a reflecting crystal must be very long in order to intercept the whole beam passed by the collimator.

4.2. Curved Crystal Spectrometers

(*a*) Cylindrically curved reflection crystals may be used in X-ray spectrometers to diffract and "focus" radiation diverging from a point or line source as shown in Fig. 4-3*a* (see also Chapter 10, Sec. 10.1). The diffracting planes must be curved to a radius equal to the diameter of the focusing circle but the surface of the crystal should be ground to lie on the radius of the circle,[1] Fig. 4-3*b*, for best resolution. If the surface is not ground,[2] the diffracted radiation will not converge to a line image (Fig. 4-3*c*).

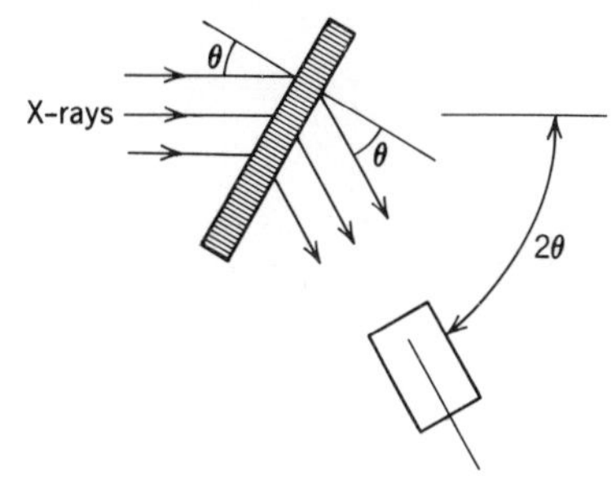

Fig. 4-2. Schematic of flat-crystal transmission optics.

A suitable source may be the focus of an electron beam as in the electron microprobe, Fig. 4-4*a*, or it may be a slit illuminated by radiation from a large specimen, Fig. 4-4*b*. In either case, the divergence corresponding to that allowed by the collimator in Sec. 4.1 above is given by the angular extent of the source as seen from the crystal. This angular divergence is negligible in the case of the electron probe but is comparable to the blade collimator in the case of the slit.

(*b*) Cylindrically curved crystals may also be used in transmission arrangements as shown in Fig. 4-5. In this case, the planes perpendicular to the crystal surface are used for diffraction rather than the planes parallel to the surface. Again the radius of curvature is equal to the diameter of the focusing circle. No grinding of the crystal is necessary however. The interesting feature of this type of crystal is that it may

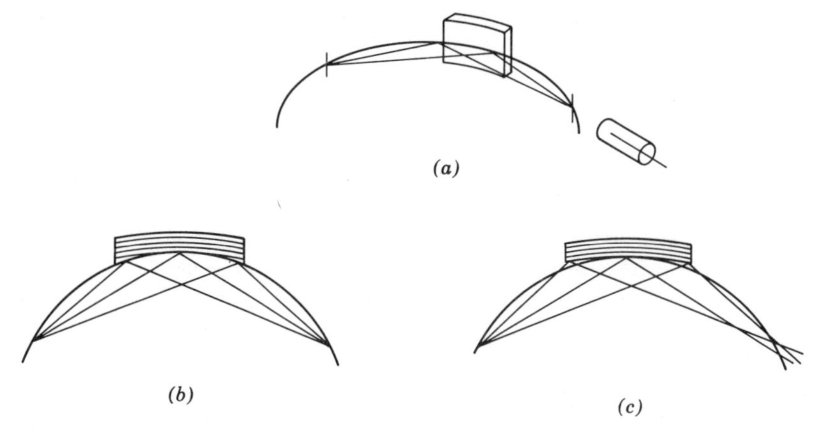

Fig. 4-3. Schematic of curved-crystal reflection optics. (*a*) Line source, crystal. line image, and detector. (*b*) Johansson curved and ground crystal for best focusing. (*c*) Johann curved-only crystal for partial focusing.

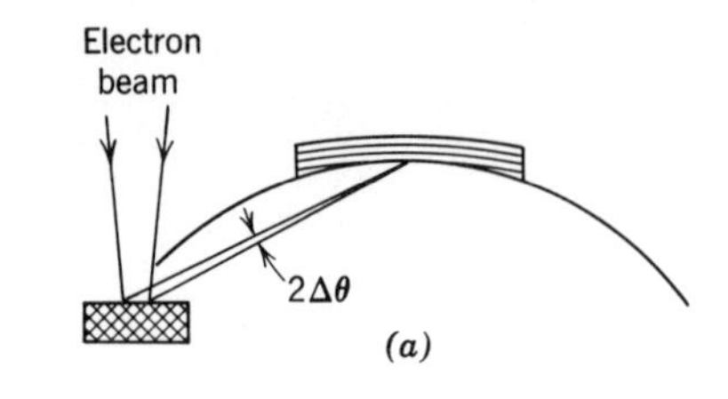

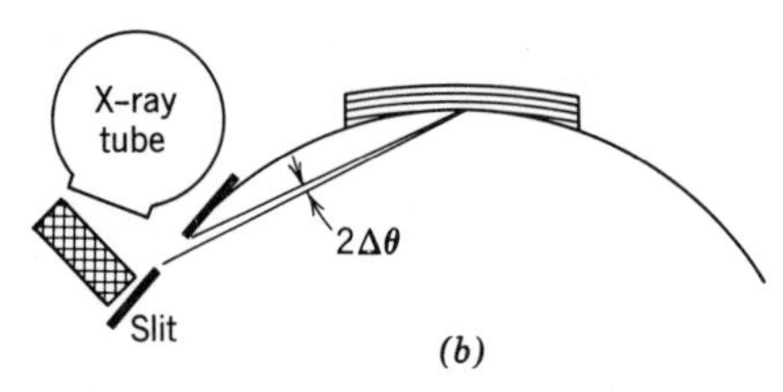

Fig. 4-4. A suitable source for curved-crystal optics may be (*a*) A focused electron beam as in the case of the electron probe or (*b*) a slit in the case of X-ray fluorescence excitation.

be used in two different ways. If a large source of fluorescent X-radiation such as a flat specimen is used in the *B* position in Fig. 4-5, the various wavelengths will be diffracted to specific points along the focusing circle. On the other hand, if a point source of radiation is located on the focusing circle at the *A* position, only one wavelength will be diffracted selectively by the crystal and the diffracted radiation will continue to diverge. This latter arrangement is used to advantage with the electron probe discussed in Chapter 10.

4.3. Other Geometries

(*a*) Doubly curved reflection crystals may be used to focus radiation diverging from a point source back to a point image provided the wavelength is just right. Referring back to Fig. 4-3, imagine the figure of revolution obtained by rotating the arc of the focusing circle about the chord joining the source and image. If the crystal is curved to fit this surface after first bending to the diameter of the focusing circle, the image will be a point rather than a line. Such double curving is possible with plastically deformed crystals but the resulting crystal will be proper for point focusing only one wavelength. For all other wavelengths, the second curvature will be of improper radius and a line image will be obtained. It is also possible to use thin slices of cylindrically curved crystals arranged on the figure of revolution surface to approxi-

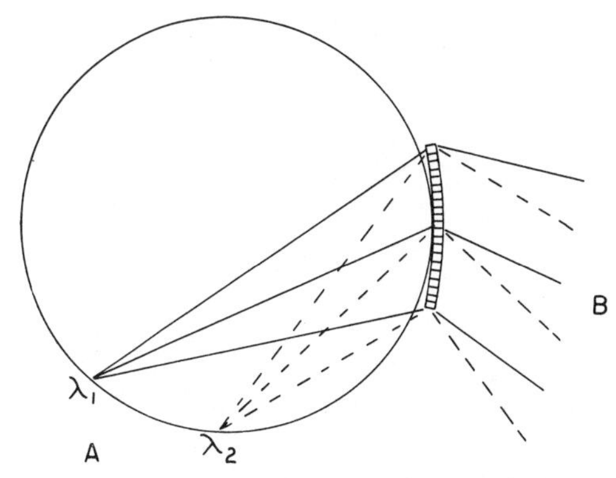

Fig. 4-5. Schematic of curved-crystal transmission optics.

mate the doubly curved crystal. The sections may be rearranged for various wavelengths, but this does not lend itself readily to scanning through a spectrum. It should be mentioned that doubly curved crystals are more valuable in X-ray diffraction where one is interested in only a single wavelength.

(*b*) Figure 4-6 shows the geometry of the small, compact, edge-crystal spectrograph.[3] The planes parallel to the thin edge are used for diffraction, and the breadth of each diffracted line is just the width of the edge projected onto the circle. Crystals as thin as 0.005 in. may be prepared from LiF and resolution of Cr K_β and Mn K_α obtained. Each wavelength diffracted arises from a different portion of the specimen; thus a homogeneous specimen is required so that each wavelength will be properly represented quantitatively. The advantage of the edge–crystal geometry is that no moving parts are required and an extremely simple spectrograph may be constructed. The complete spectrum is recorded at one time on photographic film placed along the arc of the circle and the total time for recording a complete spectrum is 10–30 min with a standard diffraction tube as the source of primary radiation.

Figure 4-7 shows the spectra from brass containing Mo in one case and plus Ti in the other case obtained simultaneously with a double spectrograph where the two specimens are placed side by side and a pie shaped blade extending from the crystal to the film prevents overlap of diffracted radiation. For quantitative results a comparison standard must be used as one of the specimens or the film sensitivity must be known as a function of wavelength. Table 4-1 gives the measured sensitivity for Eastman Kodak No-Screen X-ray film[4] for use in absolute calibration.

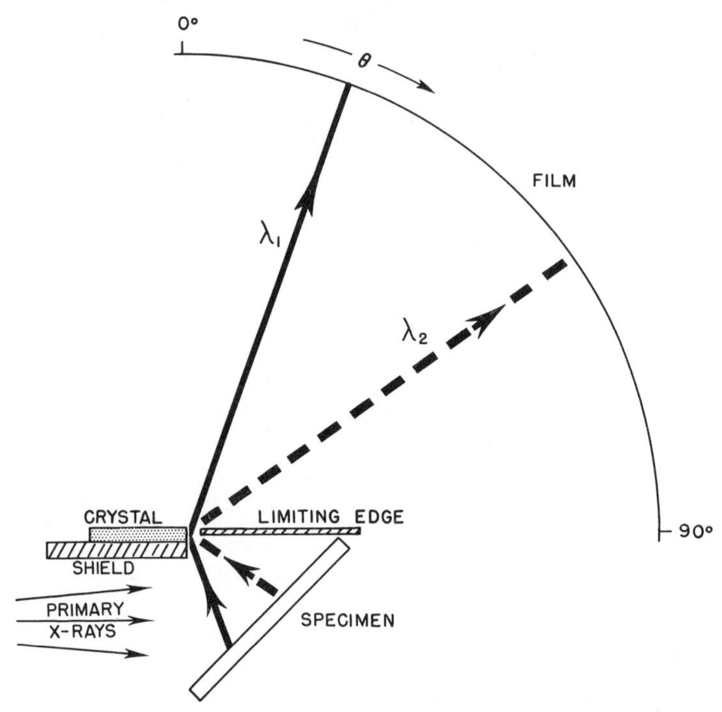

Fig. 4-6. Schematic of edge-crystal geometry for a simple X-ray spectrograph.

One primary use of the edge–crystal spectrograph is in laboratories which possess diffraction equipment and want a quick semiquantitative chemical analysis as an aid to powder diffraction identification.

(*c*) Another film-type of X-ray spectrograph uses a convex curved crystal[5] as shown in Fig. 4-8. The crystal is a thin slab (0.005 in.) of LiF curved to a radius of ⅛ in. Each wavelength from the source finds an area on the continuously curved convex surface which is at the proper Bragg angle for diffraction of that particular wavelength. Thus the complete spectrum is recorded simultaneously.

Resolution of the spectrograph depends on the angular divergence of the source and the spread of the diffracted beam by the finite thickness of the crystal. The instrument is well suited to measuring the spectra from flash X-ray tubes or hot plasmas where the time duration of the signal is too short for electronic detectors. It is also well suited for measuring the spectral distributions in regular X-ray tubes and can distinguish the presence of low concentrations of impurities in the targets. Figure 4-9 shows the spectrum from a standard Cu-target X-ray diffraction tube. The effects of the Ag and Br absorption edges in the photo-

TABLE 4-1

Sensitivity of No-Screen Film[a]

Multiply all numbers by 10^7 to obtain photons/cm^2

Density units	Energy (keV)									
	5	6	7	8	9	10	11	12	13	14
0.1	0.3	0.3	0.3	0.4	0.5	0.5	0.6	0.7	0.8	0.5
0.2	0.5	0.5	0.6	0.7	0.8	0.9	1.1	1.2	1.4	0.9
0.3	0.7	0.8	0.9	1.0	1.2	1.4	1.5	1.8	2.1	1.3
0.4	0.9	1.1	1.2	1.4	1.6	1.8	2.1	2.4	2.8	1.8
0.5	1.1	1.3	1.5	1.8	2.0	2.3	2.6	3.0	3.6	2.3
0.6	1.4	1.6	1.9	2.2	2.5	2.8	3.2	3.7	4.3	2.7
0.7	1.6	1.9	2.2	2.5	2.9	3.3	3.7	4.3	5.1	3.2
0.8	1.9	2.2	2.5	2.9	3.4	3.8	4.3	5.0	5.9	3.7
0.9	2.2	2.5	2.9	3.4	3.8	4.4	5.0	5.7	6.8	4.3
1.0	2.4	2.8	3.2	3.8	4.3	4.9	5.5	6.4	7.6	4.8
1.1	2.7	3.1	3.6	4.2	4.7	5.4	6.1	7.0	8.4	5.3
1.2	3.0	3.4	3.9	4.6	5.2	5.9	6.7	7.7	9.1	5.8
1.3	3.2	3.7	4.3	5.0	5.7	6.5	7.3	8.4	10.0	6.3
1.4	3.5	4.1	4.7	5.4	6.2	7.1	8.0	9.2	10.9	6.9
1.5	3.8	4.4	5.1	5.9	6.7	7.7	8.7	10.0	11.9	7.5
1.6	4.2	4.8	5.5	6.4	7.3	8.3	9.5	10.9	12.8	8.1
1.7	4.5	5.2	6.0	6.9	7.9	9.0	10.2	11.8	13.9	8.8
1.8	4.8	5.6	6.4	7.4	8.5	9.6	10.9	12.6	14.9	9.4
1.9	5.1	6.0	6.9	8.0	9.1	10.3	11.7	13.5	16.0	10.1
2.0	5.5	6.4	7.3	8.5	9.7	11.0	12.5	14.4	17.1	10.8
2.1	5.8	6.8	7.8	9.0	10.3	11.7	13.3	15.3	18.1	11.4
2.2	6.2	7.2	8.3	9.6	11.0	12.5	14.1	16.3	19.3	12.2
2.3	6.6	7.6	8.8	10.2	11.6	13.2	14.9	17.2	20.4	12.9
2.4	7.0	8.1	9.3	10.8	12.3	14.0	15.8	18.2	21.6	13.7
2.5	7.3	8.5	9.8	11.4	12.9	14.7	16.7	19.2	22.8	14.4
2.6	7.7	9.0	10.3	11.9	13.6	15.5	17.5	20.2	23.9	15.1
2.7	8.1	9.4	10.8	12.5	14.3	16.3	18.4	21.2	25.2	15.9
2.8	8.5	9.9	11.4	13.2	15.0	17.1	19.4	22.3	26.4	16.7
2.9	9.0	10.5	12.1	14.0	15.9	18.1	20.6	23.7	28.1	17.7
3.0	9.6	11.1	12.8	14.8	16.9	19.2	21.7	25.0	29.7	18.7
3.1	10.1	11.7	13.5	15.6	17.8	20.2	23.0	26.4	31.3	19.8
3.2	10.6	12.4	14.2	16.4	18.7	21.3	24.2	27.8	33.0	20.8
3.3	11.3	13.1	15.0	17.4	19.9	22.6	25.6	29.5	35.0	22.1
3.4	12.0	13.9	16.0	18.5	21.1	24.0	27.2	31.3	37.1	23.5
3.5	12.9	15.0	17.2	19.9	22.7	25.8	29.3	33.7	39.9	25.2
3.6	14.1	16.3	18.7	21.7	24.8	28.2	32.0	36.8	43.6	27.5
3.7	15.5	18.1	20.7	24.0	27.4	31.2	35.3	40.7	48.2	30.5
3.8	17.4	20.2	23.2	26.9	30.7	34.9	39.6	45.6	54.0	34.1
3.9	20.2	23.5	26.9	31.2	35.6	40.5	45.9	52.8	62.6	39.6
4.0	24.9	29.0	33.2	38.6	44.0	50.0	56.7	65.3	77.4	48.9

[a] Tables for energies up to 50 keV may be obtained from the authors of Ref. 4.

graphic emulsion are easily seen along with the Cu K_α and K_β lines. There are impurity lines of Cr, Fe, and Ni as shown. Figure 4-10 shows the Fe K_α and K_β lines and the ionized Fe^{+24} line from a pinch plasma with an iron anode.[6] The spectrograph may also be used with the electron probe or any other electron excited specimen. For fluorescence excited specimens the exposure is usually excessively long.

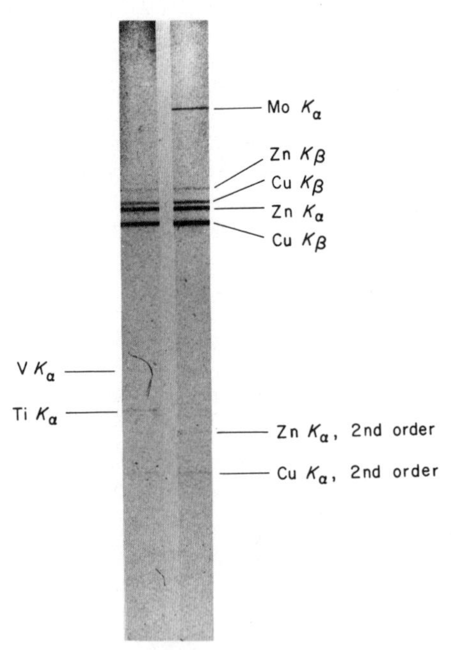

Fig. 4-7. Spectra of alloys taken with the edge-crystal spectrograph showing characteristic lines.

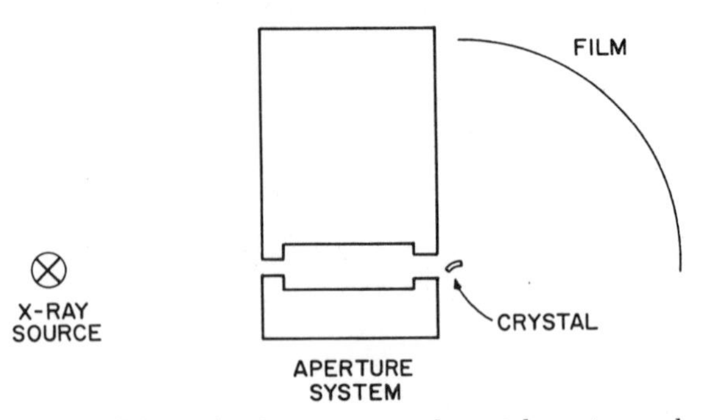

Fig. 4-8. Schematic of convex curved-crystal spectrograph.

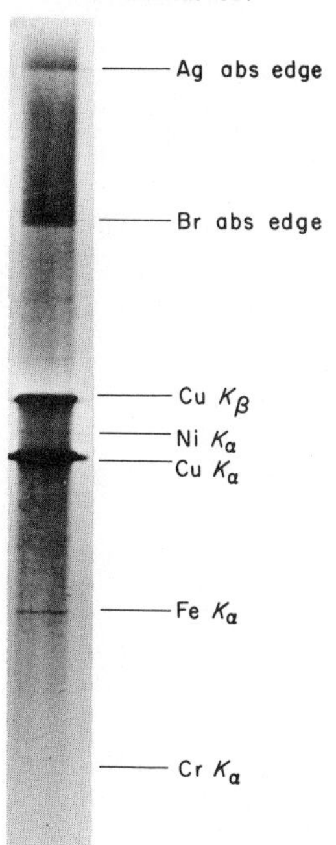

Fig. 4-9. Spectrum of a Cu target X-ray tube taken with the convex curved-crystal spectrograph showing impurity lines from Cr, Fe, and Ni.

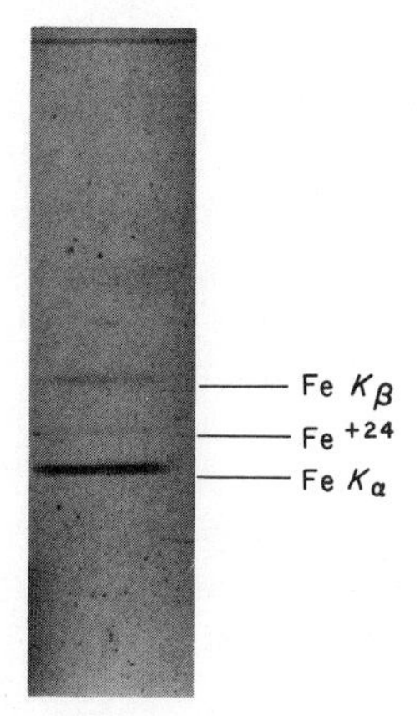

Fig. 4-10. Fe lines from a pinch plasma. In addition to the usual Fe K_α and K_β lines, the line from iron ionzied to Fe^{+24} appears.

4.4. Crystal Properties

The analyzing crystal is the heart of the X-ray spectrometer. It has four properties which determine its capabilities.

(*a*) The interplanar spacing, d, determines the angle, θ, at which the nth order of wavelength, λ, will be diffracted according to Bragg's equation, $n\lambda = 2d \sin \theta$, as was explained in Chapter 2, Sec. 2.4. Appendix 2 lists the common crystals used in X-ray spectrometers and the diffracting planes and d spacings. The maximum wavelength which can be diffracted by a crystal is given by $\lambda_{max} = 2d$ because $\sin \theta$ cannot be greater than unity. The d spacing also controls the dispersion, $\Delta\theta/\Delta\lambda$, where $\Delta\theta$ is the small change in diffracting angle for a small change, $\Delta\lambda$, in wavelength. Dispersion is found by differentiating Bragg's equation to obtain

$$\Delta\theta/\Delta\lambda = n/2d \cos \theta \tag{4-1}$$

Large spacing crystals can diffract both short and long wavelengths but the dispersion is poorer with large spacing crystals. For example, consider the dispersion in the neighborhood of Cu K_α for LiF and for KAP crystals.

$$\text{LiF: } 2d = 4.02 \text{ Å}; \theta = 22.5°; \Delta\theta/\Delta\lambda = 0.27 \text{ rad/Å}$$

$$\text{KAP: } 2d = 23 \text{ Å}; \theta = 3.8°; \Delta\theta/\Delta\lambda = 0.044 \text{ rad/Å}$$

Thus LiF gives six times better dispersion than KAP. In practice one should generally use the smallest-spacing crystal possible consistent with the maximum wavelength to be measured.

Dispersion, along with rocking curve breadth, W, discussed in paragraph *b*, below, and collimator divergence, B_c, discussed in Sec. 4.1, determines whether or not neighboring wavelengths can actually be resolved in a spectrometer. Examples are given in Sec. 4.5.

(*b*) The other crystal parameters, W, P, R, are illustrated in Fig. 4-11; they depend on the perfection of the crystal and its composition. Most crystals are made up of small mosaic blocks a few hundred angstroms in size and with slight differences in orientation. Thus a crystal does not diffract monochromatic radiation at only a single Bragg angle but, rather, over a small range of angle. To explain Fig. 4-11, imagine a strictly parallel, monochromatic beam incident on the crystal in the left-hand portion of the figure. As the crystal is rocked through the Bragg angle, θ, diffracted intensity will be spread out as shown in the right-hand portion of the figure. The breadth at half maximum, W, depends on the mosaic spread, and the peak diffraction coefficient, P,

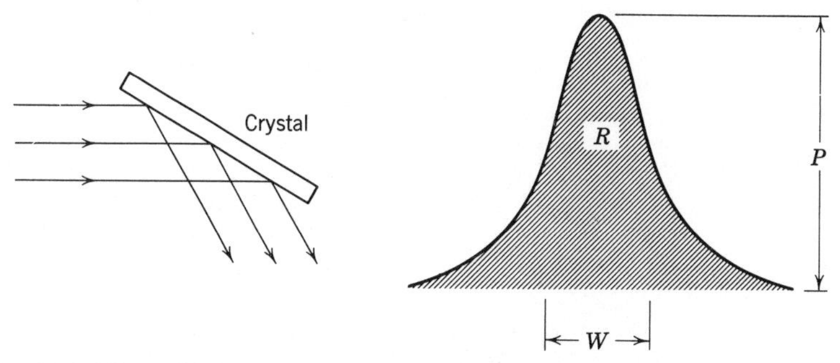

Fig. 4-11. The rocking curve which would be obtained from a real crystal diffracting parallel monochromatic radiation (rocking curves are actually measured on double crystal spectrometers).

depends to some extent on composition (through the atomic scattering factors and space group). The integral reflection coefficient, R, is the area under the curve. W is in units of radians or degrees; P is just the fraction diffracted/incident photons at the peak setting θ and is unitless; R therefore, has the same units as W, and is usually given in radians. In X-ray spectrometers the divergence allowed by the collimator is usually far greater than the rocking curve breadth of the crystal so that the whole rocking curve is included in each measurement; *therefore, two different crystals should be compared by their R values rather than by their P values.* Note that in the electron probe the source size is so small that the rocking curve breadth is the limiting factor and crystals should be compared by their P values.

Table 4-2 lists the parameters of some common crystals. The measurements were made with a double crystal spectrometer[7] because it is not possible in practice to obtain the strictly parallel monochromatic beam hypothesized in describing Fig. 4-11. The tenfold increase in R for abraded LiF[8] is most important because it means that measured intensities in X-ray spectroscopy were increased 10 times merely by treating the crystal. Almost all of the LiF crystals used in modern spectrometers are treated to gain this advantage. The increase in R for LiF is the result of increased mosaic spread caused by edge dislocations introduced by the abrading. Unfortunately most common crystals other than alkali halides cannot be improved in this simple fashion; in calcite for instance, abrading increases W but reduces P proportionally so that R remains about constant.

The favorable property of LiF led to experiments designed to maximize the number of dislocations.[9] Figure 4-12 shows a comparison of

TABLE 4-2

Parameters of Some Common Crystals[a]

Crystal	Condition	W (sec)	P (%)	R (rad)
LiF	Fresh cleavage	14	40	3×10^{-5}
LiF	Abraded	110	50	4×10^{-4}
Calcite	Cleavage	14	45	4×10^{-5}
Topaz	Ground and etched	105	7	4×10^{-5}
EDdT	Sawed	125	20	2×10^{-4}
KAP	Cleaved	70	30	5×10^{-5}

[a] All crystals except KAP were measured using Cu K_α radiation; KAP was measured using Al K_α radiation.

R vs. λ for cleaved, abraded, and flexed LiF along with theoretical curves calculated for perfect and ideally imperfect crystals.[10] Flexing was done by heating the crystal to about 400°C, bending to a radius of 20 cm and then flattening again. Thus flexing introduces dislocations throughout the crystal volume whereas abrading introduces them only to a depth of about 0.002 in. below the surface. Therefore, the flexing is most effective in improving R for the shorter wavelengths which penetrate below the abraded surface. It should be noted in Fig. 4-12 that the flexed curve is still below the theoretical imperfect crystal by about a factor of 2, which means there is still room for improvement. An interesting paper on carbon crystals[11] indicated that heating Ceylon graphite causes the layers to collapse forming a grossly mosaic crystal and giving estimated R values up to perhaps 2×10^{-3} radians at a breadth, W, of 1°. It is the low absorption coefficient and imperfections which make both LiF and C well suited to X-ray fluorescence.

In summary, it appears that the best crystals for X-ray spectrometers will be those which are nearly ideally mosaic and which consist of the lightest atoms possible because this means less absorption of the diffracted radiation. The increased R values for graphite were accompanied by an unacceptable increase in W, but eventually techniques should be available for optimizing the value of R while holding W within specified limits for adequate resolution.

4.5. Examples of Resolution Calculations

For the practicing X-ray analyst it is desirable that he be able to estimate whether or not neighboring wavelengths or overlapping orders will be resolved in his spectrometer. This aids him in interpreting data and predicting applications.

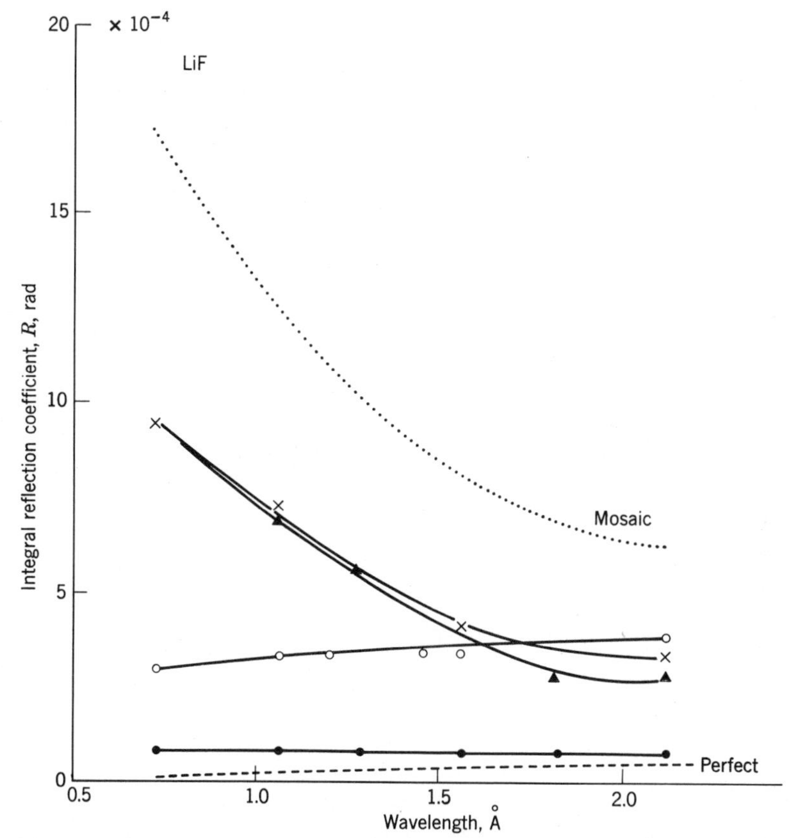

Fig. 4-12. The effect of crystal treatment on integral reflection coefficient. Theoretical values for ideally perfect and ideally mosaic crystals are shown for comparison. ··· calculated. —— measured. (●) cleaved. (○) abraded. (▲) flexed. (X) flexed and abraded.

First, we need to know the divergence, B_c, of the collimator (see Sec. 4.1). For 0.005 in. blade separation and 4 in. length

$$B_c = \arc \tan (s/l) = 0.07°$$

Next, we must know W for the crystal. For abraded LiF this was 0.033° (120 seconds of arc) in Table 4-1. The combined breadth, B, due to collimator and crystal is the convolution of the triangular intensity peak of Fig. 4-1c and the rocking curve of Fig. 4-11. B is given by

$$B^2 = W^2 + B_c^2 = 0.006 \tag{4-2}$$

$$B = 0.078°$$

Sample calculations are given below for the Cu K_{α_1} K_{α_2} doublet and the Cr K_β, Mn K_α neighboring lines.

(*a*) For Cu K_{α_1} K_{α_2} find the separation in wavelength.

$$\begin{aligned} \text{Cu } K_{\alpha_2} &= 1.5444 \text{ Å} \\ -\text{Cu } K_{\alpha_1} &= 1.5406 \\ \hline \Delta\lambda &= 0.0038 \end{aligned}$$

From Eq. 4-1 and 2-3; $\Delta\theta = (1 \times 0.0038)/(4.02 \times \cos 22°) = 0.001$ radians $= 0.057°$.

The spread of each line is given by $B = 0.078°$ in Eq. 4-2. Figure 4-13 shows the expected appearance of the K_{α_1} and K_{α_2} lines and the composite curve. Note that the K_{α_2} line is always half the intensity of K_{α_1} as was explained in Sec. 2.2. Thus from Fig. 4-13, we see that the Cu K_{α_1} K_{α_2} doublet is not resolved in the first order diffraction for the conditions specified.

(*b*) For Cr K_β Mn K_α we proceed in the same way

$$\begin{aligned} \text{Mn } K_{\alpha_1} &= 2.1018 \\ -\text{Cr } K_\beta &= 2.0849 \\ \hline \Delta\lambda &= 0.0169 \end{aligned}$$

$$\Delta\theta = 0.0049 \text{ radians} = 0.28°$$

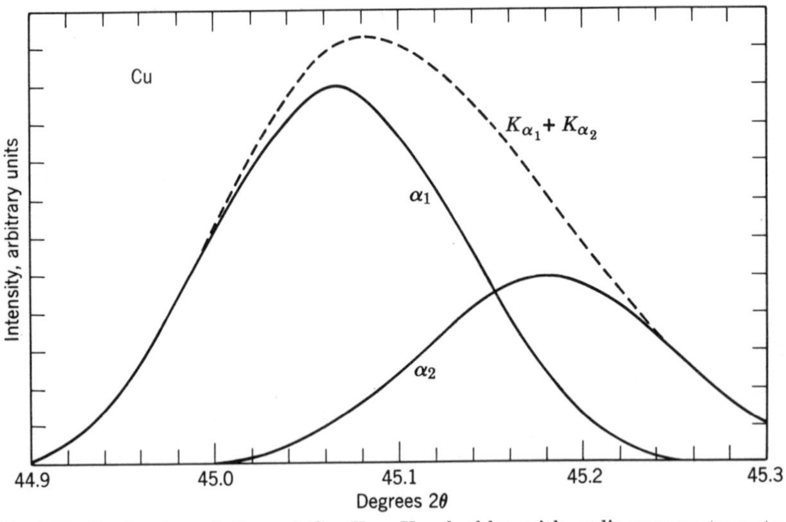

Fig 4-13. Lack of resolution of Cu $K_{\alpha1}$, $K_{\alpha2}$ doublet with ordinary spectrometer.

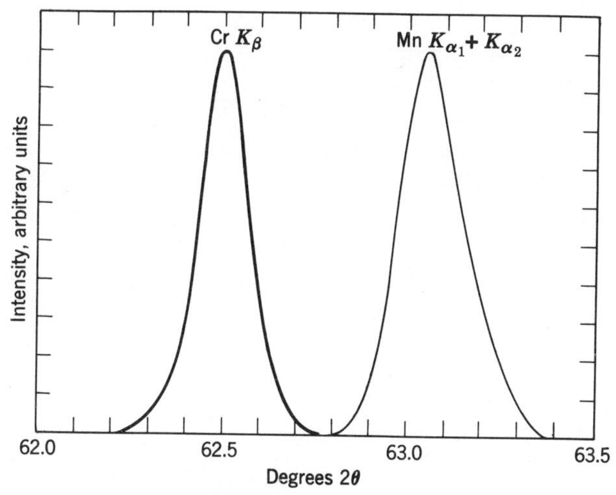

Fig. 4-14. Resolution of Cr K_β and Mn K_α doublet with ordinary spectrometer.

B is the same as before, 0.078°. Figure 4-14 shows the expected appearance. In this case resolution is expected to be complete for the Cr K_β and Mn K_α but the Mn $K_{\alpha_1}K_{\alpha_2}$ doublet is unresolved as was the Cu doublet in Fig. 4-12.

4.6. Overall Efficiency of Spectrometers

By overall efficiency of a spectrometer system we mean the fraction of characteristic X-ray photons generated in the specimen which are available to be measured by the detector:

(*a*) Consider first the fraction passed by the collimator. If the specimen of Fig. 4-1 is close to the collimator, the photons from any point on the specimen can only pass through one pair of blades. The fraction of photons generated to photons which pass through the blades is just the fractional solid angle given by the area between a pair of blades divided by the area of the sphere of radius equal to the collimator length. Referring to Fig. 4-1 for a blade separation, s, of 0.005 in. and a p value of 0.75 in. and a length, l, of 4 in., the fractional solid angle is $(0.005 \times 0.75)/(4\pi \times 4^2) = 3.7 \times 10^{-5}$.

(*b*) For a single crystal the measured fraction of incident photons diffracted at the peak position is about 5% (this is less than the values of P of nearly 50% for double crystal measurements because of the divergence of the incident radiation).

(*c*) The detector efficiency as will be discussed in Chapter 5 can be made about 90%.

The overall efficiency is just the product of *a*, *b*, and *c* above. For the examples used this becomes

$$3 \times 10^{-5} \times 0.05 > 0.90 \approx 1.7 \times 10^{-6}$$

This says that for a million characteristic photons generated in the specimen only about two will be detected. The limit set by the collimator is by far the most severe factor. Energy dispersion as discussed in Chapter 6 eliminates the collimator and crystal and increases available intensity by several hundred times because it is easy to intercept fractional solid angles larger than 10^{-3}.

References

1. T. Johansson, *Naturwissenschaften*, **20**, 758 (1932).
2. H. H. Johann, *Z. Physik*, **69**, 185 (1931).
3. L. S. Birks and E. J. Brooks, *Anal. Chem.*, **27**, 1147 (1958). L. S. Birks, U.S. Pat. 2,842,670 (1958).
4. C. M. Dozier, J. V. Gilfrich, and L. S. Birks, *Appl. Opt.*, **6**, 2136 (1967).
5. L. S. Birks, U.S. Pat. 2,835,820 (1958).
6. L. Cohen, U. Feldman, M. Sehwartz, and J. H. Underwood, *J. Opt. Soc. Am.*, **58**, 843 (1968).
7. A. H. Compton and S. K. Allison, *X-Rays in Theory and Experiment*, Van Nostrand, New York, 1935.
8. L. S. Birks and R. T. Seal, *J. Appl. Phys.*, **28**, 541 (1957).
9. J. Vierling, J. V. Gilfrich, and L. S. Birks, Submitted to *Appl. Spectry.*
10. International Tables for X-Ray Crystallography, Kynoch Press, Birmingham, England (1959).
11. R. W. Gould, S. R. Bates, and C. J. Sparks, *Appl. Spectry.*, **22**(5), 549 (1968).

CHAPTER 5

DETECTORS AND CIRCUITS

The three common detectors for X-ray spectrochemical analysis are the proportional counter (sealed or flow), the scintillation counter, and the solid-state counter. All three types operate on the principle of electronic amplification of the pulse generated each time an X-ray photon is absorbed. The principles of pulse amplitude distributions and energy discrimination are discussed in detail only for the proportional counter but are equally valid for the other detectors.

5.1. Proportional Counters

Proportional counters use a gas such as Xe, Ar, He, or CH_4 to absorb the X-rays and generate electrical pulses. Figure 5-1 shows the mechanical construction and electrical connections schematically.

(*a*) The sealed proportional counter of Fig. 5-1*a* will be discussed first. It is used for X-rays of wavelength shorter than about 2 to 3 Å. The central wire is insulated from the metal shell and maintained at a + high voltage (HV) of 1500–1800 V. The metal shell is grounded for operator protection. Entrance and exit windows are usually of mica or beryllium to transmit radiation shorter than 2 Å without appreciable absorption. The exit window is desirable so that radiation not absorbed by the gas will pass out of the counter instead of striking the metal shell which could generate spurious pulses.

When an X-ray photon enters the detector it may be absorbed in either of two ways. The least likely way is by knocking out an inner-shell electron from a gas atom which generates a fluorescent X-ray photon characteristic of the gas atom and leads to the "escape peak" phenomenon discussed in Sec. 5.5. The most likely way for the entering photon to be absorbed is by knocking out a valence electron and giving it a kinetic energy nearly equal to the photon energy. Thus a Cu K_α photon of 8000 eV energy would eject a valence electron from a Xe atom with 8000 eV minus the ionization potential of Xe. The ejected electron will lose its energy by ionizing other Xe atoms and generating a number of ion pairs, i.e., positive ions and electrons. The most probable number of ion pairs, p, is given approximately by

$$p \approx \text{X-ray energy}/(2 \times \text{ionization potential}) \qquad (5\text{-}1)$$

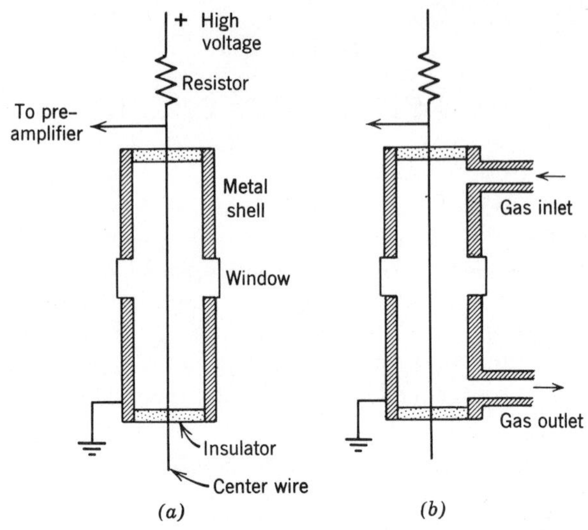

Fig. 5-1. Schematic of (*a*) sealed proportional counter and (*b*) flow proportional counter.

The factor of 2 is merely an empirical number which corrects the listed 1st ionization potential (12 eV for Xe) to an average value which agrees with observation. For the Cu K_α photon $p = 8000/(2 \times 12) \approx 330$ ion pairs.

The positive ions are attracted toward the metal shell and neutralized. The electrons are attracted toward the central wire and accelerated by the +HV so that each starting electron gains enough energy to ionize other gas atoms along the way and each of these electrons ionizes still other gas atoms until an avalanche of perhaps 10^4 electrons strikes the wire for each initial ion pair. All of this happens within about 0.1–0.2 μsec after the X-ray photon is absorbed. The many avalanches of electrons striking the wire in such a short time interval lowers the voltage on the wire momentarily and this drop in potential is transmitted as an electrical pulse through the capacitor to the amplifying circuits and read-out circuits. During the time it takes for the voltage on the wire to recover (about 1 μsec) the detector cannot respond to another X-ray photon; this is called the "dead time" of the detector. Dead time corrections to counting rates are discussed in Sec. 5.8.

Because of the linear relation between ion pairs and photon energy, Eq. 5-1, and the linear amplification by the circuits, the amplitude of the pulse observed is proportional to the X-ray photon energy—hence the name proportional counter.

(*b*) Flow proportional counters, Fig. 5-1*b*, differ from sealed proportional counters mainly in the window thickness and the filling gas. To detect X-rays in the 2–10 Å range the most common window material is quarter-mil (0.00025 in.) Mylar film and the gas is 90% Ar–10% CH_4. For the range above 10 Å the windows must be even more transparent than Mylar and common materials are stretched polypropylene or Formvar [poly(vinyl formal)]. The gas filling for the range above 10 Å may have higher proportions of CH_4 up to 100% CH_4 or may substitute He for Ar. Operating voltages will depend on the specific filling. For all the thin windows mentioned there is leakage of the gas out of the detector and air into the detector. This requires constant replenishment of the gas filling and leads to the flow principle where fresh gas is flowed into the detector constantly at a rate of 1–2 ft^3/hr.

The absorption of X-ray photons, the generation of ion pairs, and the resulting avalanches and electrical pulses are exactly the same as in the sealed counters but the ionization potential of the gas used is to be substituted in Eq. 5-1.

5.2. Pulse Amplitude Distributions

In Sec. 5.1 the most probable number of ion pairs per photon was given by Eq. 5-1; the result was an electrical pulse of amplitude proportional to the X-ray photon energy. Actually there is a statistical uncertainty in the number of ion pairs per photon so that any particular X-ray photon may generate more or less pairs than the average. Thus, for a large number of incident Cu K_α photons there is a distribution in the number of ion pairs per photon and hence in the amplitudes of the pulses recorded; the standard deviation of the distribution of ion pairs is simply $\sigma = (p)^{1/2}$.

Figure 5-2 illustrates approximately what happens when 10,000 each of Cu K_α and Fe K_α photons are absorbed in a sealed Xe proportional counter. This would be the situation with a Cu–Fe sample and the crystal spectrometer set at zero degrees so that the counter could see both Cu K_α and Fe K_α photons simultaneously. We will disregard the K_β components in the hypothesis but this does not affect the principles discussed. In Fig. 5-2*a*, the number of photons is plotted vs. photon energy and the graph consists of two "spikes" because of the monochromatic nature of characteristic lines. In Fig. 5-2*b*, the number of pulses is plotted vs. the pulse amplitude (expressed as ion pairs per pulse) and shows the expected histogram; the integral areas under the Cu K_α and Fe K_α distributions are still 10,000 pulses each. For Cu K_α,

$p_{Cu} = 8000/24 \approx 330$ and $\sigma_{Cu} = (330)^{1/2} \approx 18$. For Fe K_α, $p_{Fe} = 6400/24 \approx 270$ and $\sigma_{Fe} = (270)^{1/2} \approx 16$. In Fig. 5-2*c*, the pulses have been amplified so that the peak of the Fe distribution appears at 27 V and the peak of the Cu distribution appears at 33 V. The sum of these two distributions is approximately what one would see on a multi-

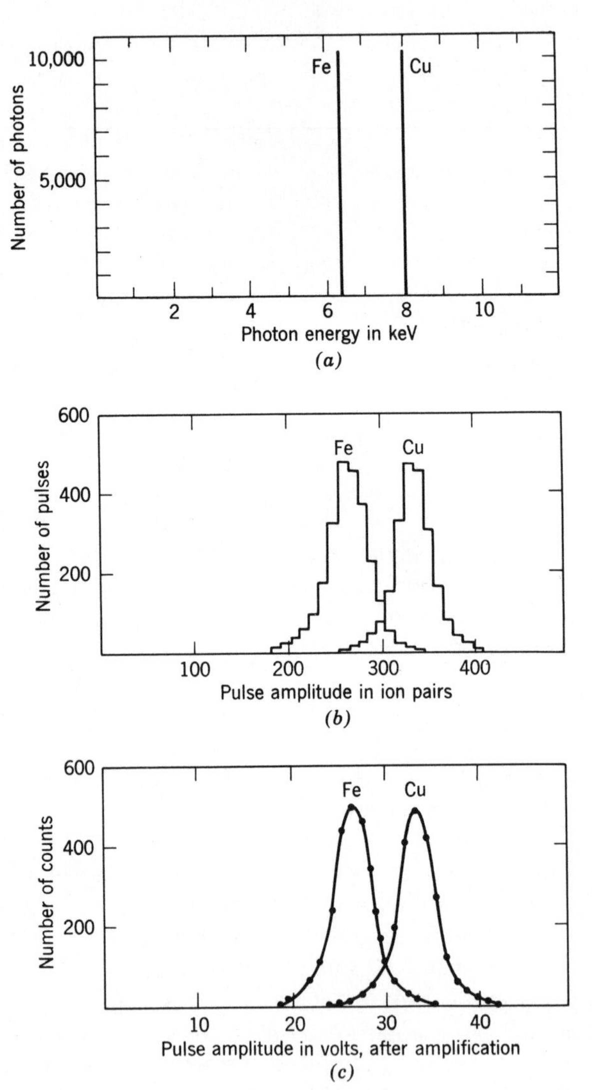

Fig. 5-2. The conversion of Cu and Fe photons to pulse amplitude distributions in the detector and amplifier.

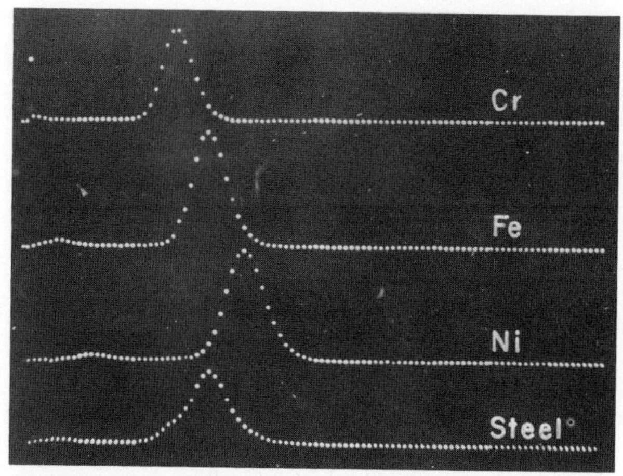

Fig. 5-3. Pulse amplitude distributions from Cr, Fe, Ni, and steel as they appear on a multichannel analyzer (proportional counter).

channel analyzer (Sec. 5.3) with the crystal spectrometer set at zero degrees as described above and with a Fe–Cu specimen.

Figure 5-3 is an actual photograph of the display tube of a multichannel analyzer showing the distributions for individual samples of Cr, Fe, and Ni and the combined distribution for a sample of stainless steel. Of course when the crystal spectrometer is used and is set for a particular element only the pulse amplitude distribution of that element (plus higher order diffraction of $\lambda/2$, $\lambda/3$, etc.) will appear, no matter how many other neighboring elements are present in the specimen.

5.3. Energy Discrimination and Multichannel Analyzers

Electronic circuits may be set to pass only pulses larger than or smaller than some desired amplitude. For instance, in Fig. 5-2*c* an electronic gate could be set to pass only pulses larger than 24 V. This would pass most of the Fe pulses and all of the Cu pulses. If a second gate were then set to pass only pulses smaller than 30 V, the band pass would be 24–30 V which would still pass most of the Fe pulses but would discriminate against most of the Cu pulses. Such gating circuits called pulse height analyzers are commonly used in X-ray fluorescence analysis with crystal spectrometers; they are set to pass all of the radiation, λ, for which the spectrometer is set but will discriminate against second-order diffraction of $\lambda/2$, third-order diffraction of $\lambda/3$, etc. The lower gate is usually called the "base line"; the upper gate or "window" setting is at the desired number of volts above the base line. The same circuits are also used in X-ray diffraction to eliminate higher order diffraction of portions of the continuum.

The multichannel analyzer operates on a somewhat different principle. The amplitude of each pulse from the detector is read electronically and the pulse is stored in the appropriate memory channel. After a pre-set time, the number of pulses stored in each channel is displayed as was shown in Fig. 5-3. Usually there are two or four sets of 100 or 200 channels each so that several readings of different specimens may be made sequentially and then displayed simultaneously. The range of pulse amplitudes stored in a set of channels is controlled simply by adjusting the gain of the input amplifier. What one sees on the display tube is the energy spectrum of the X-rays entering the detector but blurred by the spread of the pulse amplitude distribution for each energy. The multichannel analyzer is, by far, the fastest way to examine a range of X-ray energies because all amplitudes are accepted at the same time. More will be said about the multichannel analyzer in Chapter 6 on energy dispersion (nondispersive analysis).

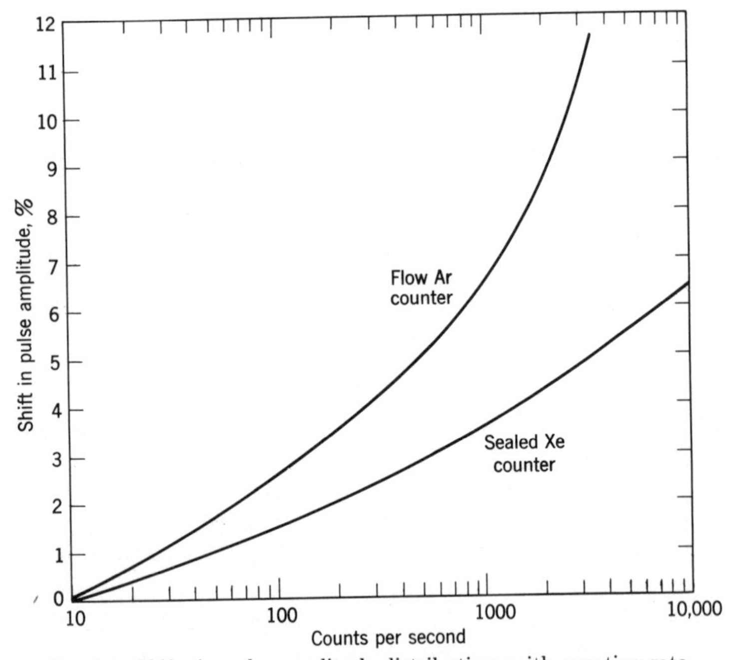

Fig. 5-4. Shifts in pulse amplitude distributions with counting rate.

5.4. Shift of Pulse Amplitude Distributions with Counting Rate

When counting rates exceed a few thousand counts per second, the pulse amplitude distribution begins to broaden and to shift to lower amplitude. There are at least three possible causes for this: (*a*) the positive ion sheath moving toward the shell reduces the effective accelerating potential for electrons moving toward the wire and amounts to the same thing as reducing the +HV on the wire; (*b*) the number of electrons per avalanche decreases to some extent at high counting rates; (*c*) particles of dirt or roughness on the central wire distort the local electric field and broaden the pulse amplitude distribution. Shifts are noticeably worse in flow counters than in sealed counters. Figure 5-4 shows how large the shifts may be in the peaks of the distributions in standard flow and sealed counters. The best way to reduce the shift to a minimum is to operate the counter at the low end of the +HV range (it will be necessary to increase the gain of the amplifier to bring the distribution back to the desired value). Usually the settings for the base line and window can be set for a wide enough band pass to include any shift in the distribution. The optimum settings are best determined experimentally for the particular counter and for the counting rate limits to be encountered.

5.5. Escape Peak

In Sec. 5.1, the least likely way for an incident X-ray photon to be absorbed was by knocking out an inner-shell electron in a gas atom. This *K* or *L* shell ionization does happen at a certain fraction of the time, however, and leads to the escape peak phenomenon which is described as follows: The characteristic X-ray photon emitted by the gas atom in this process is likely to escape from the detector because the gas has a low absorption coefficient for its own radiation. The remainder of the incident photon energy is available in the form of kinetic energy of electrons and generates a number of ion pairs according to Eq. 5-1. These ion pairs lead to a second pulse amplitude distribution at a reduced energy. For instance, for incident Cu K_α photons and an argon–methane flow counter the 8047 eV Cu K_α energy is reduced by 2957 eV corresponding to Ar K_α emission, leaving 5090 eV which will generate an average of 212 ion pairs. The lower energy or escape peak is shown in Fig. 5-5 to the left of the main Cu peak. In sealed Xe counters it is usually the Xe *L* escape peak which is observed because most incident photons do not have the 34 keV required to *K* ionize Xe.

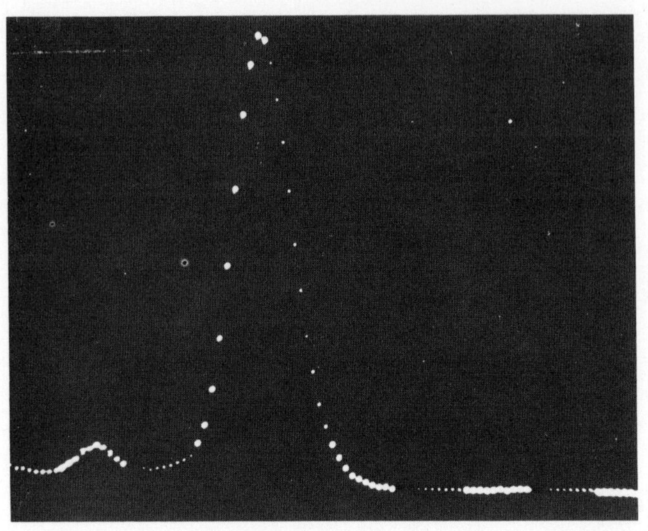

Fig. 5-5. Main pulse amplitude distributions for Cu, right, and escape peak, left. (Argon–methane dector)

Pulses appearing in the escape peak must be added to the main peak for proper representation of the incident X-ray intensity. This is done by setting the base line and window of the circuits to include both peaks. The effects of escape peaks on energy dispersion analysis are discussed further in Chapter 6.

5.6. Scintillation Counters

The scintillation counters used for X-ray analysis consist of a thallium activated NaI crystal sealed to the window of a photomultiplier tube to amplify the signal generated. Figure 5-6 shows the construction schematically. When an X-ray photon is absorbed in the crystal it generates a number of visible-light photons (scintillations) instead of ion pairs as was done in the gas proportional counter. The visible-light photons strike the light sensitive surface of the photomultiplier and cause electrons to be emitted. The group of electrons emitted for each X-ray photon are multiplied in number by about 10^6 times by knocking out an increased number of electrons at each accelerating grid. Thus, a large pulse of electrons appears at the output of the photomultiplier for each incident X-ray photon. The number of visible-light photons per X-ray photon is larger than the number of ion pairs in a proportional counter because the energy per event is only 2–3 eV instead of 20–30 eV

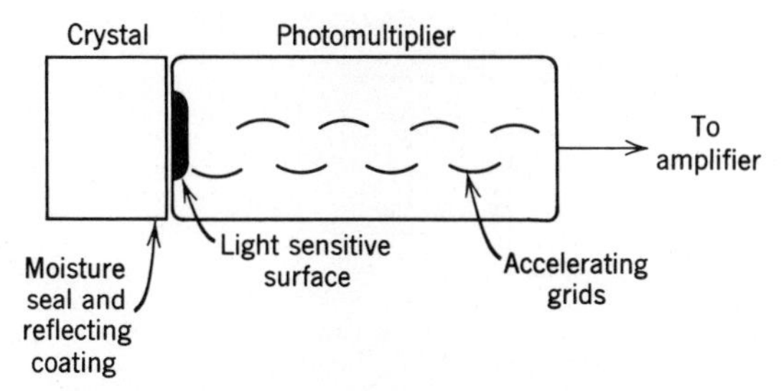

Fig. 5-6. Schematic of scintillator–photomultiplier detector.

per ion pair. Again there is a statistical spread in the number of events per X-ray photon and one might expect $\sigma\%$ to be relatively less because of the larger number of events (visible photons). However, only about 1 out of 10 or 20 visible photons results in an electron being emitted from the light sensitive surface of the photomultiplier so that the effective number of events per X-ray photon is actually less than in a gas proportional counter. Therefore the $\sigma\%$ is greater and the pulse amplitude distribution broader by as much as two or three times. Figure 5-9 in Sec. 5.8 shows comparative breadths of pulse amplitude distributions for proportional, scintillation, and solid-state counters. When the scintillation counter is used with a crystal spectrometer it has adequate resolution to discriminate against higher-order diffraction.

The principal advantage of the scintillation counter is its ability to completely absorb incident X-rays of the highest energy used in X-ray analysis, i.e., up to 50 keV. It is appreciably more efficient than the gas proportional counter for all energies above about 6 keV; for X-ray photons below 6 keV it is difficult to measure the photon pulses above the spurious noise pulses from the photocathode. One interesting combination detector is a gas proportional counter ahead of a scintillation counter. Long wavelength X-rays which would not pass through the moisture-proof layer needed to protect the NaI crystal are absorbed and measured by the proportional counter. Shorter wavelengths which pass through the proportional counter are measured by the scintillation counter. Both signals are fed into the same amplifying system and recorded in the usual way.

5.7. Solid-State Counters

The newest type of X-ray detector is single-crystal semiconductor material such as Si or Ge. Figure 5-7 shows the detector schematically.

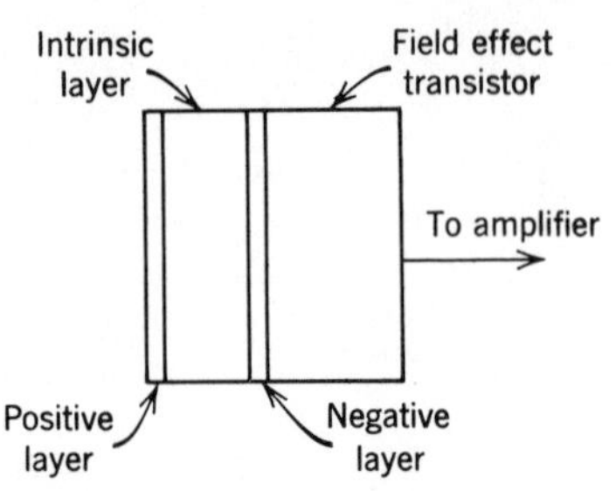

Fig. 5-7. Schematic of solid-state detector and attached field-effect transistor.

The Si has low concentrations of other elements added in layers to make the surface *n*-type (negative), the middle layer *i*-type (intrinsic or neutral), and the back layer *p*-type (positive). When a voltage of 300–900 V is applied as shown, no current should flow in the detector. If an X-ray photon is absorbed in the middle layer, it generates a number of electron–positive hole pairs. The electrons and holes have mobility in the Si and are quickly drawn to the front and back surfaces, respectively, resulting in an electrical pulse. The average energy required to generate an electron-hole pair is about 3.8 eV for Si corresponding to a large number of events per X-ray photon and resulting in a small value for the relative statistical spread of the pulse amplitude distribution. This means better resolution than either the gas proportional or scintillation counters. The theoretical resolution for silicon detectors (about $2\sigma\% = 5\%$ or 300 eV for Mn K_{α}) was limited originally by electronic and thermal noise but is being approached in newer commercial equipment. In practice, the resolution depends on the volume and shape of the detector and on the bonding to the amplifier stages. Even a resolution of 400 eV is only achieved presently by cryogenic cooling; at room temperature the statistical spread would correspond to X-ray photon energies of several keV and would mean extremely broad pulse amplitude distributions.

Operation at cryogenic temperatures usually means vacuum operation in order to prevent condensation of moisture. In a vacuum spectrometer at cryogenic temperature the solid-state detector can probably be used to advantage for wavelengths as long as 5–6 Å before the 400 eV resolution is poorer than that of gas proportional counters.

5.8. Comparison of Detectors

There are several ways to compare detectors such as efficiency, resolution, speed (dead time), or restrictions-on-use.

(*a*) Efficiency: By efficiency we mean the fraction or percentage of the incident beam which will be absorbed in the active volume of the detector. Figure 5-8 shows the efficiencies for proportional, scintillation, and solid-state detectors. The efficiency takes account of the usual size of detector (1 in. diameter for Ar and Xe proportional counters, 0.5 in. for scintillation crystal, and 1 mm zone thickness for Si and Ge solid-state detectors) and type of window or absorbing layer ahead of the active volume.

(*b*) Resolution: The spread in the pulse amplitude distributions has been discussed in previous sections. Figure 5-9 shows a comparison for Ti, Cu, and Mo radiation. The proportional counter is always better than the scintillation counter but the solid-state detector is better than either of the others for wavelengths up to a few angstroms. For wavelengths greater than 10 Å the proportional counter would be expected to have better resolution than the solid-state detector ca. 1968 because of the electronic noise in the semiconductor material.

(*c*) Speed (dead time): The recovery time for all three types of detectors is more than fast enough so that dead-time losses are low for counting rates up to 10,000–30,000 cps. For accurate quantitative analysis, dead-time corrections should be made for counting rates above

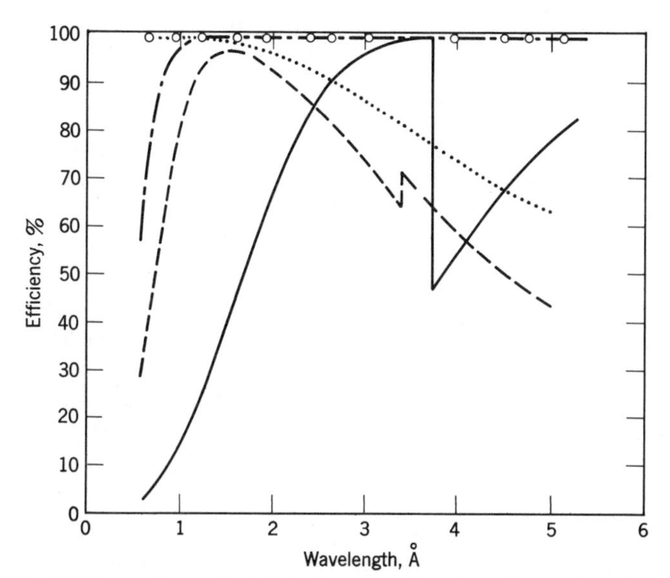

Fig. 5-8. Efficiency for various detectors as a function of wavelength. —— Ar flow proportional (¼ mil Mylar). - - - Xe sealed proportional 1 mg/cm² mica). ··· NaI scintillation (1 mil Mylar). — – — Solid state Si. ○ Solid state Ge.

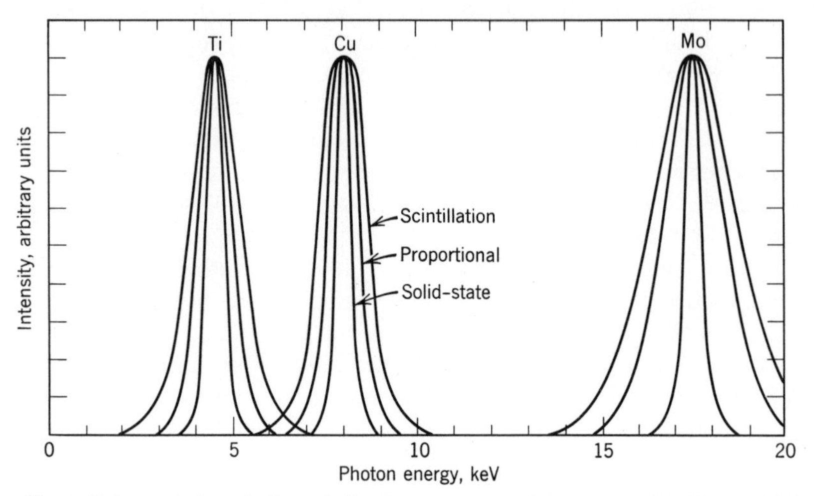

Fig. 5-9. Expected resolution of the three common detectors at Ti, Cu, and Mo wavelengths.

a few thousand counts per second. If the counting rate observed is 10,000 cps and the dead time for the detector-circuit combination is 3 μsec the amount of time in each second during which the counter is dead is simply $10^4 \times 3 \times 10^{-6} = 3 \times 10^{-2}$ sec. This is 3% of the time and means that, on the average, 3% of the incident photons were missed; the corrected counting rate becomes $10^4/0.97 = 10.3 \times 10^4$ cps. Generally speaking, counting rates should not exceed values which would require dead-time corrections of more than 10% if the simple formula above is to be used.

(*d*) Restrictions: Flow proportional counters require pressure regulators and sometimes temperature regulators to keep the gas density and hence the X-ray absorption constant enough for exacting work; for much routine analysis, however, the temperature regulator may be dispensed with especially if standards are measured frequently for comparison with the unknowns. The ultra-thin windows on flow counters for X-ray wavelengths longer than 10 Å are very fragile and usually must be replaced every few hours of operation; this is more of an inconvenience than a restriction.

Present solid-state detectors must be operated at low (cryogenic) temperatures in order to reduce electronic noise but the restriction may be overcome if other means such as gamma irradiation are effective in immobilizing excess carriers in undoped sliicon. There may be an interim period when thermoelectric cooling is adequate and this will be much more convenient than present cryogenic cooling. Solid-state detectors are easily damaged by sudden changes in bias voltage or temperature.

In scintillation counters the photomultiplier tube is permanently damaged by applying too high a voltage and the crystal is damaged if moisture penetrates the moisture-sealing layer. Other than these limitations the scintillation detector is quite rugged.

CHAPTER 6

ENERGY DISPERSION

In Chapter 4 the crystal spectrometer was used to separate the characteristic X-ray lines according to their wavelength. But in Chapter 5 it was shown that the detectors have the capability of distinguishing the different characteristic radiations by their energies. Since wavelength and energy are equally good representations of the characteristic radiations it is feasible to eliminate the crystal spectrometer and measure the emission directly as shown schematically in Fig. 6-1.

6.1. Comparisons of Wavelength and Energy Dispersion

The problems limiting practical analysis by energy dispersion have been poor resolution and high counting rates. For instance, the usual crystal spectrometer has resolution of about 0.1–0.5° θ depending on the collimator. This corresponds to a resolution of $\Delta\lambda/\lambda = \cot\theta\Delta\theta$ of 0.4 to 2.0% at Cu K_α. The best resolution with solid-state detectors is about 400 eV which corresponds to about 5% at Cu K_α. Figure 6-2 shows comparative resolution for spectrometers and detectors over the range of interest. Usually the spectrometers will resolve adequately the lines of interest, even the K_α line of element Z from the K_β line of element $Z - 1$ as well as those L lines of higher atomic number elements which overlap the K lines of lower atomic number elements. Conversely, the detectors alone will not resolve the K_α line of element Z from the K_β line of $Z - 1$ at all and will not completely separate the K_α lines of Z and $Z \pm 1$ except for solid-state detectors at $Z > 20$. However, mathematical unfolding (Sec. 6.2) of overlapping lines can often be used to separate the components, and therefore energy dispersion becomes practical even for neighboring elements when a precision of 3–4% can be tolerated. Precision approaches that of wavelength dispersion for systems of elements separated by a few atomic numbers.

High counting rates attained by eliminating the crystal spectrometer would seem intuitively to be an advantage but this is not necessarily true. The counting rate limitations and shift of pulse amplitude distributions discussed in Chapter 5 apply to the total signal entering the detectors. In the case of the crystal spectrometer the line from the element of interest comprises nearly all of the total signal, but in energy

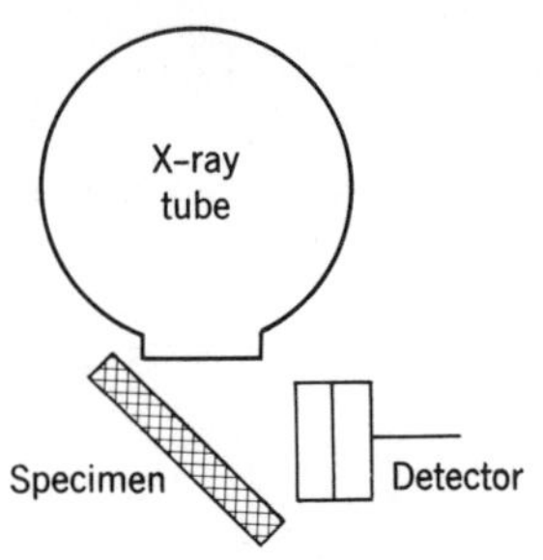

Fig. 6-1. Schematic of energy dispersion using only an X-ray tube, specimen, and detector.

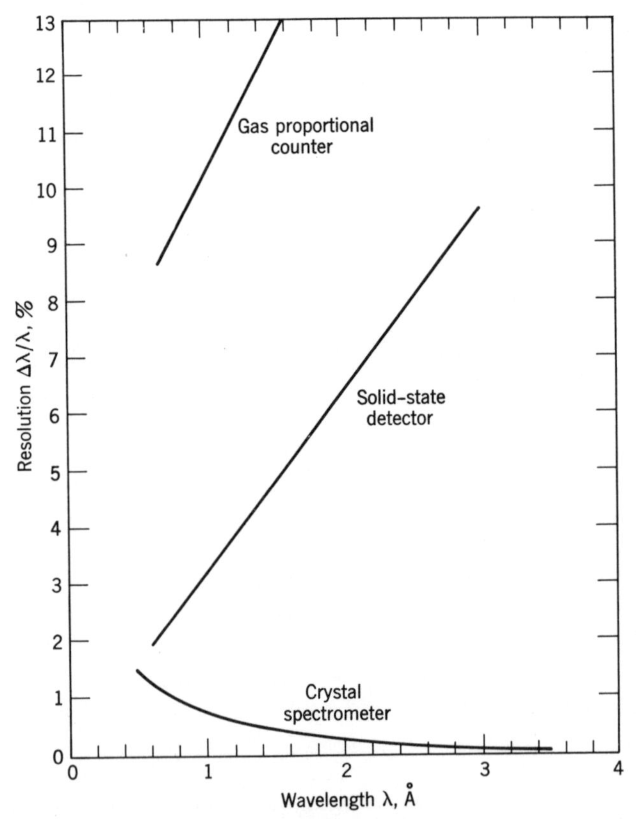

Fig. 6-2. Resolution, $\Delta\lambda/\lambda$, for proportional and solid-state detectors used for energy dispersion compared with resolution of ordinary crystal spectrometer.

dispersion it comprises only part (sometimes a small part) of the total signal. Thus the usable counting rate for the element of interest will often be lower in energy dispersion than in wavelength dispersion. In addition, shifts in the pulse amplitude distributions at high counting rates are difficult to treat in the mathematical unfolding technique and therefore limit the precision of the analysis. Electronic techniques are available to prevent a pulse–amplitude shift at high counting rates by automatically adjusting the detector voltage; such techniques are not recommended for energy dispersion of multicomponent systems because the amount of shifting required is not the same for all elements. The increased efficiency of energy dispersion does allow practical use of radioisotope sources as discussed in Sec. 6.4.

We may summarize the situation for wavelength and energy dispersion as follows: Wavelength dispersion with crystal spectrometers is required for general analytical laboratories where a variety of specimens must be accommodated and precise quantitative analysis must be achieved. Energy dispersion is advantageous for quick survey work and for repetitious quantitative analysis of systems where its resolution is known to be adequate or where mathematical unfolding can separate the component radiations. Examples are plating thickness measurement and on-stream analysis. Energy dispersion makes far more efficient use of source strength because the detector intercepts a 300 times larger solid angle of radiation from the specimen than does the spectrometer collimator. This means that compact, radioactive sources or low-power X-ray tubes have sufficient intensity for excitation and this lends itself to portable field operations or isolated equipment.

6.2. Mathematical Unfolding

Figure 5-3 of Chapter 5 showed the overlapping nature of the Cr, Fe, and Ni lines in stainless steel as measured with a proportional counter and multichannel analyzer. In order to determine their separate intensities it is necessary to use a technique such as mathematical unfolding.[1,2] Simply stated, this means to subtract the contributions from Fe and Ni at the Cr position, etc. Figure 6-3 shows how total intensity at point A is made up of the full intensity of element A plus fractional contributions from B and C. Mathematically this is expressed as

$$I_{\mathrm{MA}} = I_{\mathrm{AA}} + I_{\mathrm{BA}} + I_{\mathrm{CA}}$$

We shall now introduce the term relative X-ray intensity, that is, the intensity I_{MA} from element A in some matrix M divided by the intensity

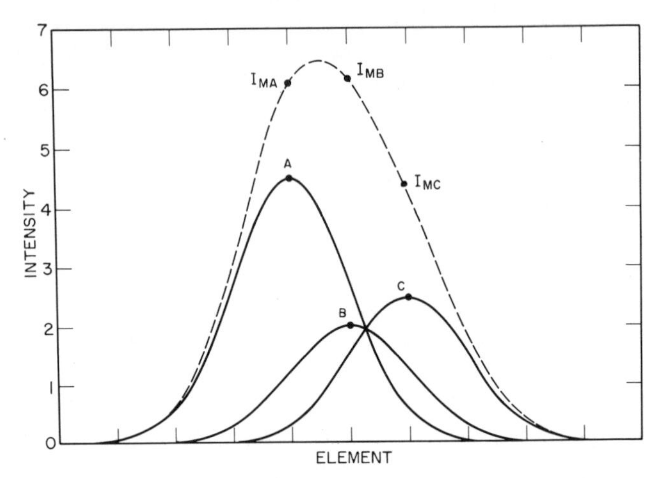

Fig. 6-3. Composite energy spectrum from three separate pulse amplitude distributions.

I_{AA} from a 100% sample of element A; this will be called $R_A = I_{MA}/I_{AA}$. It must be kept in mind that $I_{AA} \neq I_{BB} \neq I_{CC}$ because of the change in detector sensitivity with energy. In order to set up the required equations one must first measure experimentally, at constant excitation and time, the peak values of I_{AA}, I_{BB}, I_{CC} and the fractional contributions at each position, $\mathcal{A}_{AA}$, $\mathcal{A}_{AB}$, $\mathcal{A}_{AC}$, $\mathcal{A}_{BA}$, $\mathcal{A}_{BB}$, etc.; $\mathcal{A}_{AA} = \mathcal{A}_{BB} = \mathcal{A}_{CC} = 1$. Then one may write

$$
\begin{aligned}
I_{MA} &= \mathcal{A}_{AA}R_AI_{AA} + \mathcal{A}_{BA}R_BI_{BB} + \mathcal{A}_{CA}R_CI_{CC} \\
I_{MB} &= \mathcal{A}_{AB}R_AI_{AA} + \mathcal{A}_{BB}R_BI_{BB} + \mathcal{A}_{CB}R_CI_{CC} \\
I_{MC} &= \mathcal{A}_{AC}R_AI_{AA} + \mathcal{A}_{BC}R_BI_{BB} + \mathcal{A}_{CC}R_CI_{CC}
\end{aligned}
\qquad (6\text{-}1)
$$

In these equations the only unknowns are R_A, R_B, and R_C; all the other terms are measured experimentally.

As an example of the technique consider the stainless steel sample used to obtain Fig. 5-3. The printout of the multichannel analyzer for the three standards and the unknown[2] is shown in Fig. 6-4. From Fig. 6-4 the values of the constants for Eq. 6-1 are determined and shown in Table 6-1. Solution of the equations yields $R_{Cr} = 0.234$, $R_{Fe} = 0.622$, $R_{Ni} = 0.052$. These values are just as valid as if they had been obtained with a crystal spectrometer. Precision will be discussed in general in Chapter 7. When the intensities come from mathematical unfolding it is more difficult to calculate the precision than it is for

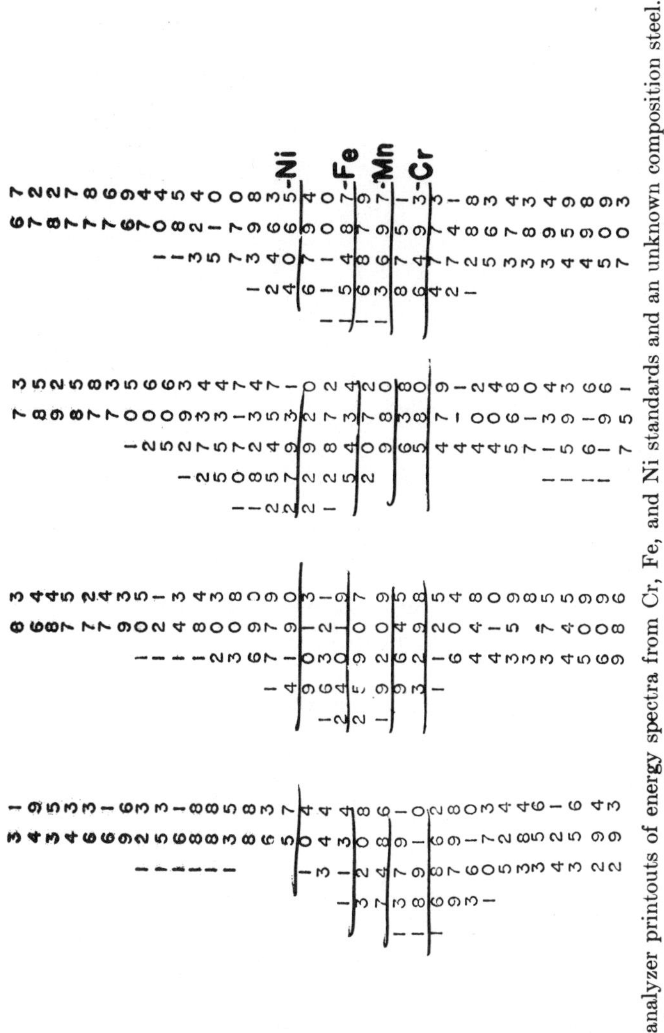

Fig. 6-4. Multichannel analyzer printouts of energy spectra from Cr, Fe, and Ni standards and an unknown composition steel.

TABLE 6-1

Values for Eq. 6-1 Obtained from Fig. 6-4

A = Cr B = Fe C = Ni

$I_{MA} = 6493$		$I_{AA} = 18910$
$I_{MB} = 15487$		$I_{BB} = 24019$
$I_{MC} = 4065$		$I_{CC} = 27931$
$\alpha_{AA} = 1$	$\alpha_{BA} = 0.1359$	$\alpha_{CA} = 0.0208$
$\alpha_{AB} = 0.0600$	$\alpha_{BB} = 1$	$\alpha_{CB} = 0.1946$
$\alpha_{AC} = 0.0030$	$\alpha_{BC} = 0.1744$	$\alpha_{CC} = 1$

resolved lines and it is usually done with a computer program. Even the unfolding must be done with a computer program when the number of components (and equations) is four or more.

6.3. Variations in Energy Dispersion Technique

Part of the limitation by resolution and intensity may be overcome by using balanced (Ross) filters or by being very selective in exciting the characteristic radiation. For instance, a 1st fluorescer may be used as in Fig. 6-5*a* to excite only some of the components. If Fe is used as the 1st fluorescer its radiation will excite Cr in steel but will not excite Fe or Ni. The sensitivity for Cr is doubled by this approach and low concentrations can be measured readily. On the other hand, if one wishes to select the high Z component, say Ni in steel, one can excite the specimen directly and then use an Fe 2nd fluorescer as in Fig. 6-5*b*. In this case the Fe intensity from the 2nd fluorescer is excited only by the Ni radiation from the specimen and can be used as a sensitive measure of the Ni content. The most complex situation arises when one wishes to measure only a middle Z element in the matrix, say Fe in steel. Figure 6-5*c* shows a Ni 1st fluorescer which will excite both Fe and Cr radiation and a Cr 3rd fluorescer which will be excited only by the Fe radiation.

All of the above schemes have been tried and give sufficient intensity for energy dispersion analysis. They are only feasible for repetitious analysis of a particular type of specimen and the fluorescers must be chosen for the particular element to be measured. For suitable applications they allow a very simple geometry and sensitive response.

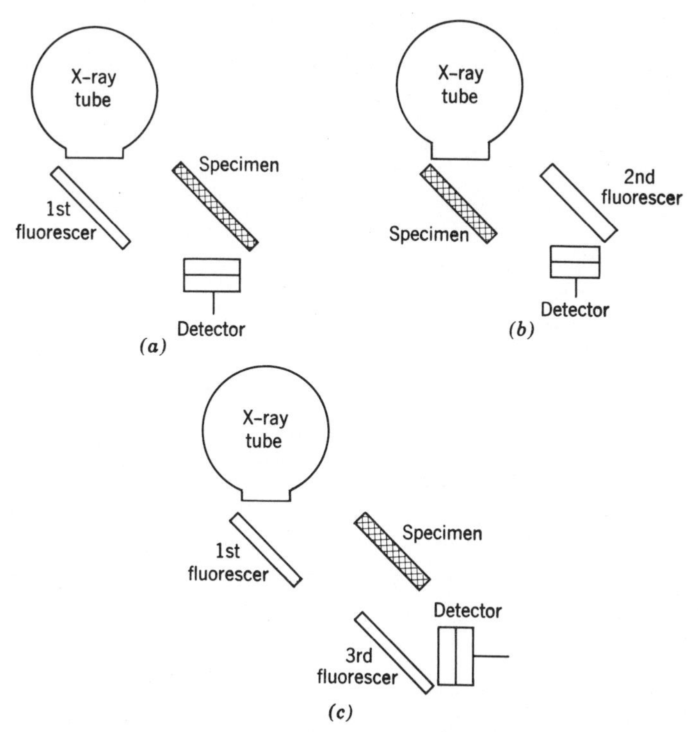

Fig. 6-5. Variations in energy dispersion geometry. (*a*) A 1st fluorescer is used to excite only elements of lower *Z* than itself in this specimen. (*b*) Only specimen elements of *Z* greater than the 2nd fluorescer will excite it. (*c*) A combination of (*a*) and (*b*) allows selection of only intermediate *Z* elements in the specimen.

6.4. Radioactive Sources for Energy Dispersion

The increased efficiency of energy dispersion compared to wavelength dispersion makes it feasible to use radioactive sources for excitation. Rhodes[3] has tabulated the properties of many of the isotope sources and his table is repeated here as Table 6-2. ^{241}Am or ^{3}H + Zr are suitable for a variety of elements while others such as ^{109}Cd are ideal for only a few elements. With the small size of the radioactive source, it is convenient to mount it as shown in Fig. 6-6. The radioactive isotope is on the annular ring around the detectors for compact geometry.

Frequently it is advantageous to use selective filters to limit the characteristic radiation reaching the detector so that mathematical unfolding of the energy spectrum is simplified or even eliminated entirely.

TABLE 6-2
Commonly Used Low Energy X-Ray and γ-Ray Sources

Source	Half-life	Useful radiations	Practical emission efficiency, photons per disintegration	Typical activity	Highest atomic number usefully excited, K X-rays
Iron-55	2.7 years	Manganese K X-rays, 5.9 keV	0.15	2 mC	24 (chromium)
Tritium–zirconium[a]	12.3 years	Bremsstrahlung, 2–12 keV	4×10^{-5}	units of 1–3 C	30 (zinc)
		Zirconium L X-rays, 2 keV	10^{-5}–10^{-4}		
Cadmium-109[b]	1.3 years	Silver K X-rays, 22 keV	0.8	1 mC	43 (technetium)
		γ ray, 88 keV	0.04		
Promethium-147–aluminium[b]	2.6 years	Bremsstrahlung, 10–100 keV	2×10^{-3}	0.5 C	60 (neodymium)
Americium-241[b]	470 years	γ ray, 59.6 keV	0.35	1 mC	69 (thulium)
		γ ray, 26 keV	0.02		
		Neptunium L X-rays, 11 to 22 keV	0–0.2[c]		
Gadolinium-153	236 days	γ ray, 103 keV	0.2	1 mC	88 (radium)
		γ ray, 97 keV	0.2		
		Europium K X-rays, 42 keV[c]			
Cobalt-57[b]	270 days	γ ray, 136 keV	0.10	0.5 mC	98 (californium)
		γ ray, 122 keV	0.88		
		γ ray, 14 keV	0–0.06[c]		
		Iron K X-rays, 6.4 keV[c]			

[a] Tritium–titanium also available.
[b] Also used as primary source in source–target assemblies.
[c] Emission depends on self-absorption of source and on window thickness.

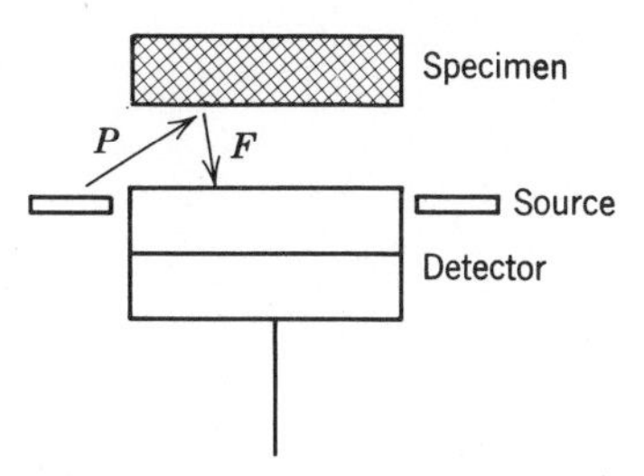

Fig. 6-6. Compact energy dispersion arrangement with a ring-shaped radioactive source shielded from the detector but irradiating the specimen; P is the primary radiation, F is the fluorescent radiation.

Recent improvements[4] in rolling thin filter layers and molding filter material in plastic have made the preparation of filters for a variety of elements much more practical than heretofore. Nevertheless the difficulties of achieving uniformity in filter thickness should be a caution to the novice.

6.5. Examples of Neighboring-Element Analysis

Several systems of nearest- and next-nearest neighbors were investigated in 1965 to determine what precision could be expected in X-ray intensities when unfolding of the energy spectra (Sec. 6.2) was necessary. One of the systems was stainless steel containing the next-nearest neighbors Cr, Fe, and Ni as major constituents and Mn as a minor constituent. An X-ray tube was used for direct excitation as in Fig. 6-1 or a Zn 1st fluorescer was used as in Fig. 6-5*a*. The energy spectra were recorded with a flow proportional counter and multichannel analyzer. Figure 5-3 showed a typical display and Fig. 6-4 showed the multichannel analyzer printout. Equations 6-1 were used in a computer program to calculate the relative X-ray intensities for the Cr, Fe, and Ni and occasionally Mn. Results were scaled to a type 316 steel standard and compared with chemical analysis in Tables 6-3 and 6-4.

In addition to the stainless steel, samples of high-temperature alloys and nickel–silver were also examined. All of these alloys contain three or more immediate neighbors which is the most difficult unfolding situation. Again the results were used to estimate composition and compared with chemical analysis as shown in Table 6-5. Note that the X-ray results are normalized to 100% whereas the constituents analyzed in 1187 and 1191 alloys add to only 90%; this contributes to the errors but is probably realistic in the sense that the analyst would ordinarily not know what to allow for the other constituents.

TABLE 6-3

Direct Excitation by X-Ray Tube
Av. of 9 runs each (neglecting Mn)

Steel	Element and wt. %	Meas. R_i (%)	Est. wt. %[a]	Difference X-ray – chem.
301	Cr 17.9	26.9	18.3	0.4
	Fe 72.7	60.3	70.8	−1.9
	Ni 7.2	4.4	7.9	0.7
303	Cr 17.2	25.4	17.3	0.1
	Fe 71.2	60.0	70.3	−0.9
	Ni 8.7	5.1	9.4	0.7
304	Cr 18.6	27.5	18.7	0.1
	Fe 69.5	58.2	68.2	−1.3
	Ni 9.4	5.4	9.6	0.2
321	Cr 17.8	26.4	18.0	0.2
	Fe 68.2	58.1	68.0	−0.2
	Ni 10.8	6.1	11.0	0.2
347	Cr 17.7	26.3	17.9	0.2
	Fe 67.9	57.1	67.1	−0.8
	Ni 10.7	6.2	11.2	0.5

[a] No. 316 steel used as the only standard for all other steels. Av. $\sigma_{obs}/\sigma_{exp} = 1.19$.

TABLE 6-4

Direct Excitation by X-Ray Tube
Av. of 10 runs each (including Mn)

Steel	Element and wt. %	Meas. R_i (%)	Est. wt. %[a]	Difference X-ray – chem.
301	Cr 17.9	27.2	18.6	0.7
	Fe 72.7	60.4	70.9	−0.3
	Ni 7.2	4.3	7.7	0.5
303	Cr 17.2	26.5	18.3	1.0
	Fe 71.2	60.2	70.7	−0.4
	Ni 8.7	5.11	9.1	0.4
304	Cr 18.6	28.7	19.4	1.2
	Fe 69.5	58.5	68.5	−0.1
	Ni 9.4	5.3	9.4	−0.1
321	Cr 17.8	27.6	18.9	0.1
	Fe 68.2	58.3	68.3	0.1
	Ni 10.8	6.0	10.7	0.0
347	Cr 17.7	26.5	18.3	0.6
	Fe 67.9	57.2	67.2	−0.6
	Ni 10.7	6.2	10.9	0.1

[a] No. 316 steel used as the only standard for all other steels. Av. $\sigma_{obs}/\sigma_{exp} = 1.33$.

TABLE 6-5

Excitation by First Fluorescer
Av. of 10 runs each, nearest-neighbor elements

Specimen & 1st fluor.	Element and wt. %	R_{meas}/R_{calcd}	wt. % est.	Difference X-ray − chem.
Alloy 1187	Cr 21.6	1.04	23.6	2.0
(Zn fluor)	Fe 27.4	0.99	29.5	2.1
	Co 20.8	0.90	20.7	−0.1
	Ni 20.3	1.15	26.2	5.9
Alloy 1191	Cr 19.5	0.93	19.0	−0.5
(Zn fluor)	Fe 2.0	1.22	2.9	0.9
	Co 13.7	0.68	10.3	−3.4
	Ni 55.2	1.07	68.2	13.0
Nickel	Ni 11.8		13.2	1.4
Silver	Cu 58.6		58.8	0.2
(Br fluor)	Zn 29.1		28.0	−1.1

Av. $\sigma_{obs}/\sigma_{expt}$ = 0.90.

TABLE 6-6

Direct Excitation by X-Ray Tube
Different Data Treatment Methods; Av. of 10 runs each

Steel	Element and wt. %	Wt. % difference; X-ray − chem.		
		Unfolding	Subtract Fe	Direct ratio
	Cr 17.9	0.4	0.5	0.7
301	Fe 72.7	−1.9	−2.5	−2.5
	Ni 7.2	0.6	0.6	3.9
	Cr 17.2	0.1	0.1	0.6
303	Fe 71.2	−0.9	−1.6	−1.6
	Ni 8.7	0.7	0.6	3.0
	Cr 18.6	0.1	0.2	0.1
304	Fe 69.5	−1.3	−1.5	−1.5
	Ni 9.4	0.2	0.2	2.3
	Cr 17.8	0.2	0.2	0.4
321	Fe 68.2	−0.2	−0.3	−0.3
	Ni 10.8	0.2	0.2	1.5
	Cr 17.7	0.2	0.2	0.3
347	Fe 67.9	−0.8	−1.0	−1.0
	Ni 10.7	0.5	0.6	1.7
	Av. error in % of amount present			
	Cr	1.3	1.4	2.4
	Fe	1.4	2.0	2.0
	Ni	5.0	4.9	27.6

When semiquantitative analysis is adequate such as in well-logging, prospecting, or moon analysis, it is possible to simplify the data treatment considerably and still achieve useful results. Figure 6-7*b* and *c* show two degrees of simplification in addition to the unfolding represented by Fig. 6-7*a*. In Fig. 6-7*b* the only correction made on the Cr intensity was to subtract the contribution of the major constituent as represented by the peak of the distribution (in this case Fe). Such a subtraction can be done with pencil and paper. In Fig. 6-7*c* the intensity at each element position is treated as being contributed by only that element even though this is obviously incorrect. Results of the simplified treatments are given in Table 6-6. Subtracting for only the major constituent gives results nearly as good as complete unfolding. No correction at all gives results which would be suitable for semiquantitative analysis.

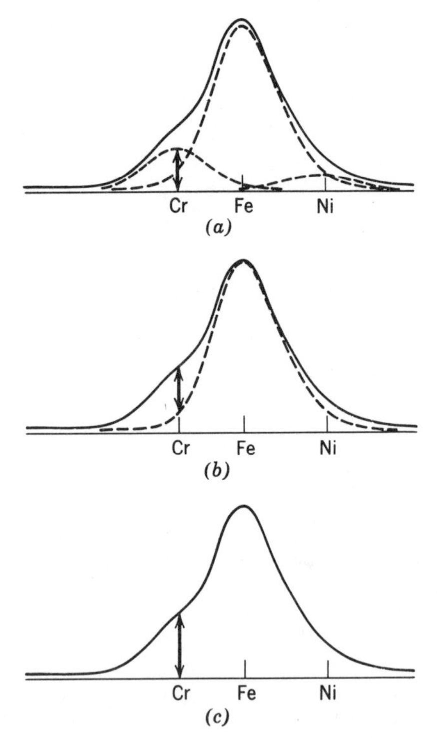

Fig. 6-7. Standard and simplified data treatment schemes for energy dispersion analysis of neighboring elements. (*a*) Unfolding simultaneous equations. (*b*) Subtract major element. (*c*) Uncorrected intensity.

As detector resolution improves with the solid-state detectors and as counting rate capabilities of multichannel analyzers increase, it is likely that more and more applications of energy dispersion will become practical.

References

1. R. M. Dolby, *X-Ray Optics and Microanalysis*, H. H. Patee, V. E. Cosslett, and A. Engstron, Eds., Academic Press, New York, 1963, p. 483.
2. L. S. Birks, R. J. Labrie, and J. W. Criss, *Anal. Chem.*, **38**, 701 (1965).
3. J. R. Rhodes, *Analyst*, **91**, 683 (1966).
4. F. O. Halliday, A. K. Kearst, J. F. Kelman, and T. O. Passell, *J. Appl. Phys.*, **38**, 1874 (1967).

CHAPTER 7

ANALYSIS, PRECISION, AND ACCURACY

All of the preceding chapters have been concerned with the physics and mechanics of generating and measuring characteristic X-ray spectra. Now we are ready to begin being concerned with translating the measured X-ray intensity into chemical composition for quantitative analysis and the estimation of precision and accuracy of the analysis. The chapter will cover the preparation of calibration curves, the use of standards, and precision. Chapter 8 will discuss two mathematical methods designed to eliminate standards as much as possible in quantitative analysis.

7.1. Calibration Curves

For quantitative analysis it is necessary to relate measured X-ray intensity to weight percent composition. Intensity is not usually linear with composition because of matrix absorption and secondary fluorescence (Sec. 7.2). Often calibration curves are used as illustrated in Fig. 7-1. Intensity may be plotted as total measured intensity, I_L, at the line peak or as line intensity above background, I_{L-B}, or as normalized (relative) intensity above background, R_i. By R_i we mean the intensity from element i in a sample of intermediate composition divided by the intensity from a 100% sample of element i (high purity is not required for the 100% sample; usually 98 or 99% is adequate). R is the most useful representation of intensity for elements where it is easy to obtain pure-element standards. For some elements such as Na, N, U, etc. this is not feasible and one must normalize against a known-composition compound or go back to the un-normalized intensity, I.

Calibration curves are obtained by measuring a range of known-composition standards or by calculation (Chapter 8). For multicomponent samples, Fig. 7-1c, the intensity vs. composition for element i is really a family of curves, A, B, C, depending on the relative amounts of the other elements present.

7.2. Matrix Effects

There are two kinds of matrix effects, namely, absorption and enhancement (secondary fluorescence). Absorption was explained in Chapter 2 and tabulations of mass absorption coefficients are given in

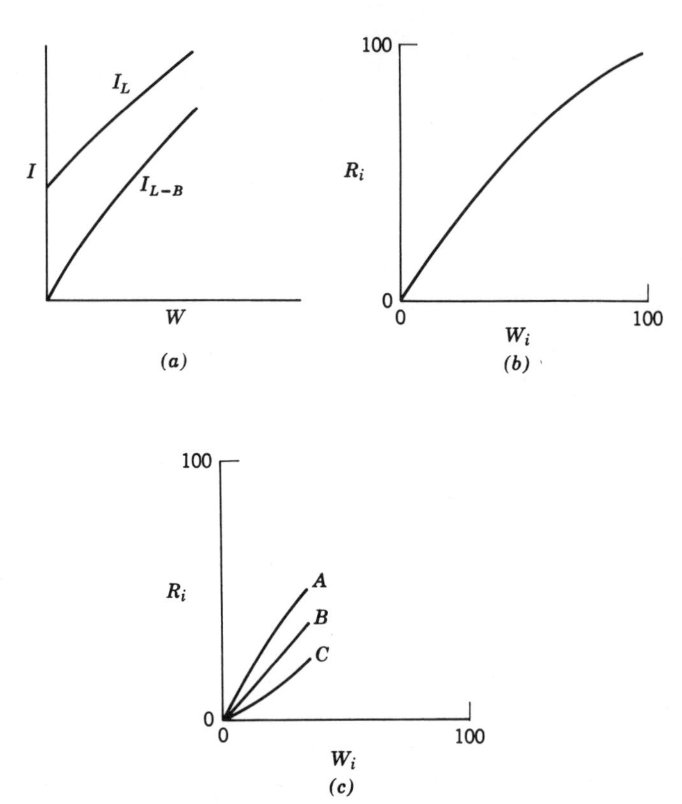

Fig. 7-1. Calibration curves. (*a*) Total line intensity or line minus background. (*b*) Relative intensity. (*c*) Multicomponent systems with a variety of matrix compositions *A*, *B*, or *C*.

the literature.[1] Enhancement occurs when the characteristic radiation from one element has enough energy to excite the characteristic radiation of one or more other elements in the sample.

The matrix processes are illustrated schematically in Fig. 7-2. Primary radiation, P_λ, of wavelength λ penetrating the specimen is absorbed by the various elements in the specimen; the total mass absorption coefficient of the matrix for wavelength λ is just the linear sum of the individual absorption coefficients times the weight fraction concentration, W_i

$$\mu_{M\lambda} = \mu_{1\lambda}W_1 + \mu_{2\lambda}W_2 + \mu_{3\lambda}W_3 + \cdots \qquad (7\text{-}1)$$

In each layer, dx, the primary radiation will excite characteristic fluorescent radiation which is emitted with equal intensity in all directions.

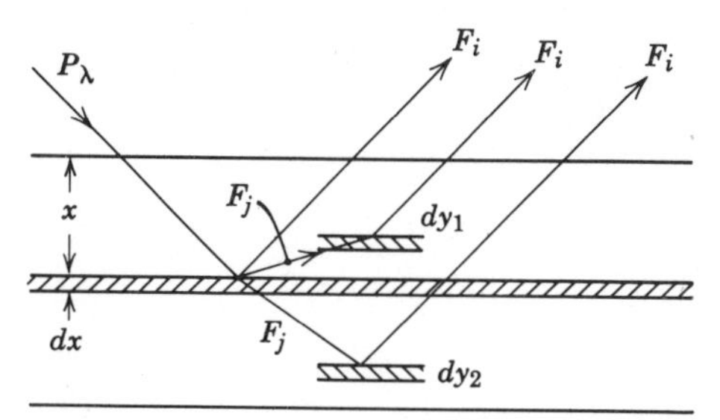

Fig. 7-2. Primary fluorescence of element j in layer dx may excite secondary fluorescence of element i in other layers such as dy_1 and dy_2.

Part of the fluorescent radiation, F_i, from element i will emerge to the surface; the mass absorption coefficient, μ_{Mi}, for the fluorescent radiation is given by

$$\mu_{Mi} = \mu_{1i}W_1 + \mu_{2i}W_2 + \mu_{3i}W_3 + \dots \qquad (7\text{-}2)$$

In addition, fluorescent radiation, F_j, from element j may excite more radiation in element i and part of this radiation will also emerge to the surface as shown. The combination of absorption of primary and fluorescent radiation and the enhancement by other elements is what controls the shape of the calibration curves.

As an example of absorption effects, consider the calibration curves for Ni combined with each of the individual elements listed in Table 7-1. The mass absorption coefficient of each element for Ni K_α radiation is given in the table. Mo with a coefficient of 220 and Cr with a coefficient of 300, as compared to the Ni self-absorption coefficient of 72, will absorb the Ni K_α strongly and depress the calibration curve as shown in Fig. 7-3. Al has almost the same absorption coefficient for Ni K_α as Ni itself but the calibration curve is well above the relation because Al has a low-absorption coefficient for the primary radiation which excites more characteristic radiation in the Ni than it would in a more absorbing matrix. C has a very low absorption coefficient for Ni K_α and the calibration curve is raised considerably. Since neither C nor Al radiations have enough energy to excite Ni by secondary fluorescence, and since Mo radiation is relatively not very effective in exciting Ni (see Fig. 3-7 on x, e, p excitation in Chapter 3), there are no noticeable enhancement effects.

As an example of enhancement, consider the calibration curves for Cr combined with Fe or Ni. Both Fe and Ni radiation can excite Cr but Fe is more effective than Ni because it is closer to the Cr absorption

TABLE 7-1
Mass Absorption Coefficients for Ni K_α Radiation

Element	$\mu_{i\text{Ni}}$
Mo	220
Ni	72
Cr	300
Al	74
C	7

edge. The overall effect on Cr is nearly identical, as shown in Fig. 7-4 because the W_{L_α} excites the Ni more strongly than the Fe. If there were no enhancement, the slightly higher absorption coefficients of Fe or Ni for Cr K_α would depress the Cr calibration curves slightly. But because of the enhancement the actual calibration curves are well above the linear relationship to composition as shown in Fig. 7-4. It should be noted that the solid curves are not simply bowed upwards as was the case for a low absorption coefficient in Fig. 7-3, but instead,

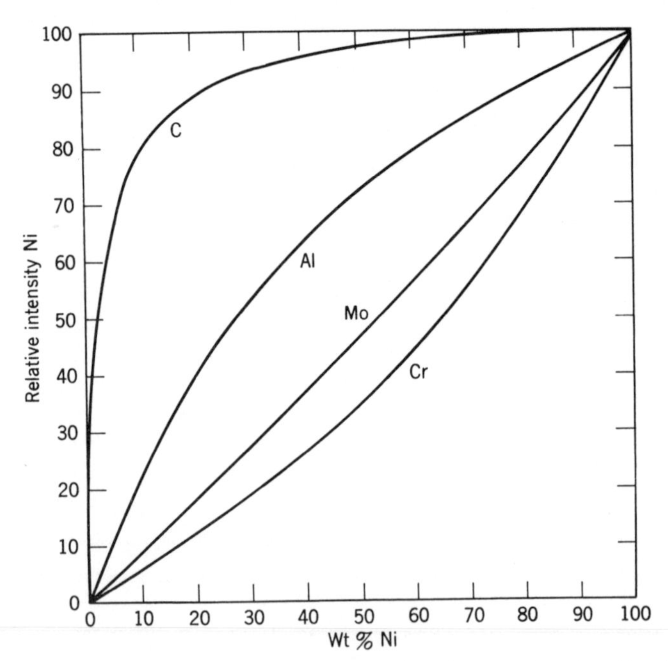

Fig. 7-3. Calibration curves for Ni in various other elements showing the effects of matrix absorption.

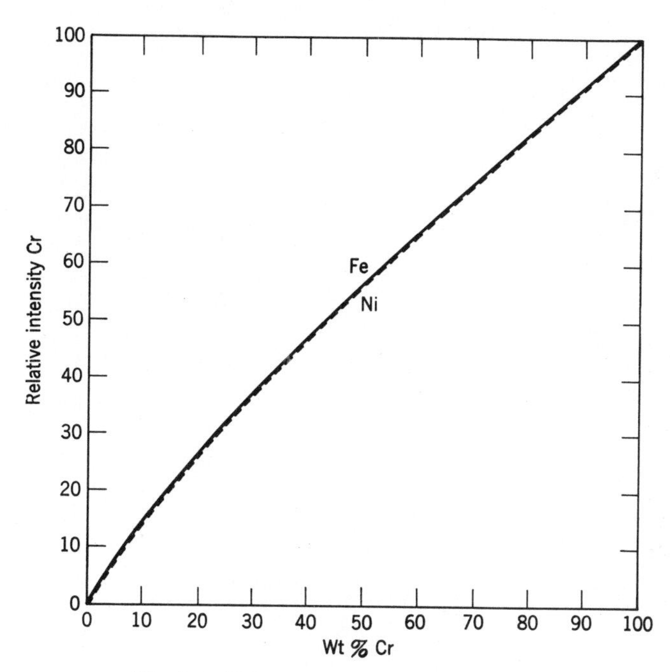

Fig. 7-4. Enhancement of Cr by Fe or Ni.

they have an inflection with a slightly reversed curvature at the higher concentrations. The significance of this is discussed in Sec. 8.1 of Chapter 8 on empirical coefficients for mathematical analysis.

7.3. Calibration of Liquids or Powders

For liquid samples it is practicable to prepare calibration curves or families of curves by making a range of known composition and measuring the resulting X-ray intensities of the elements of interest. Figure 7-5 shows such a calibration curve prepared by E. L. Gunn[2] for Zn as zinc naphthenate in oil. The calibration curves should be rechecked occasionally (usually once a day) to make sure that the equipment is operating properly.

Some powder samples may be analyzed by comparison with prepared powder standards of known composition. However, many of the powder samples are minerals or other materials with considerable variation in matrix composition. It is commonplace to dilute and fuse such samples in borax[3] to form a homogeneous solid solution and thus to reduce the matrix variation. The procedure recommended by

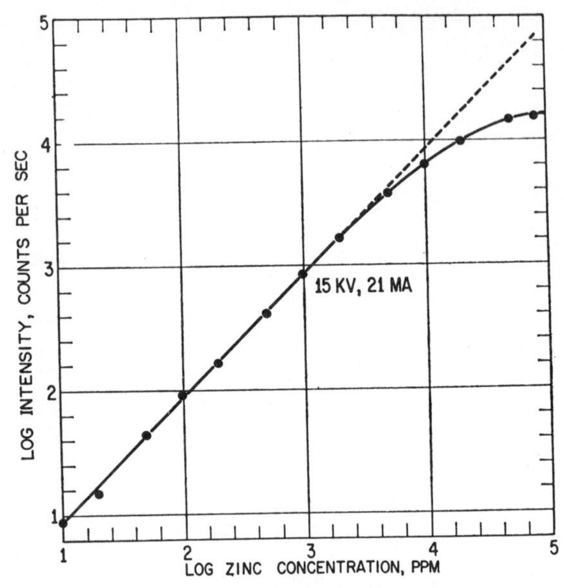

Fig. 7-5. Calibration curve for Zn as zinc naphthenate in oil (after Gunn, Ref. 2).

Claisse[3] is as follows: Place 100 mg of sample in a platinum crucible and add 10 g of borax. The crucible is heated over a flame and the sample melted; agitation hastens homogeneous solution (5–15 min). Cast the melt into a wire ring on a heated aluminum surface. Cool slowly to room temperature. The result is a smooth glass disk of uniform color ready to be examined in the X-ray spectrometer.

It is also easy to add internal standards to liquids or powders when a wide variety of compositions (such as in mineral analysis) precludes easy preparation of calibration curves. There are several requirements for internal standards:

(*1*) Internal standards are best suited to the measurement of elements that are present as less than about 10% composition although they may be used for higher concentrations on some occasions. The reason for the 10% nominal limit is that the internal standard should be added in about the same amount as the element to be determined. When more than 10% is added, it may alter the effective matrix and introduce errors in the determination.

(*2*) The internal standard element should be close to the desired element in atomic number, subject to certain limitations. If Z is the atomic number of the element to be determined, elements $Z \pm 1$ will have nearly the same absorption and enhancement coefficients with respect to the matrix. Thus a given measured intensity will correspond to the same composition of the standard element and the desired element. Caution should be used in adding standards two or three atomic numbers away from the desired element because of the likelihood of selective absorption or enhancement between the standard and the desired element. For instance, rhodium is not a suitable standard for molybdenum because the Rh K_α radiation excites the molybdenum strongly. Likewise, yttrium is not a suitable standard because it absorbs the Mo K_α radiation strongly. Some workers have used as internal standards, elements whose L radiation is close to the K radiation of the desired element in wavelength. This is not a good choice if it can be avoided because the absorption properties of the standard will be different from those of the desired element, and also because the L line intensities do not have the same relation to composition as do the K lines (L lines are only about $\frac{1}{3}$ to $\frac{1}{5}$ as strong as K lines at the same wavelength).

(*3*) Perhaps most important, the internal standard must be made homogeneous in the specimen. This may be difficult, or in some cases impossible, to accomplish if particle size is important and cannot be reduced sufficiently by grinding.

When a suitable element j is added as a standard for element i the concentration, W_i, of element i in the diluted specimen is simply

$$W_i/W_j = I_i/I_j \tag{7-3}$$

provided that the same series lines, both K or both L, are measured for i and j. If the K series of one element and the L series of the other are used, an empirical correction must be made.

Sometimes it may happen that there is no suitable element available that will serve as a good internal standard. Then the best technique is to add a known amount of the same element that is to be determined. This method only works well when the element to be determined has a composition of less than 5% or perhaps 10% because the calibration curve can only be assumed to be linear near the origin. For instance, if one is trying to determine Fe in mineral ores containing Fe_2O_3 and CuO plus other compounds, there is no suitable element that may be added whose K radiation will not be selectively absorbed or enhanced by the Fe or Cu radiation. However, a known amount of Fe_2O_3 may be added without changing the matrix effects on the Fe radiation. In Fig. 7-6 suppose that 130 counts per second were measured for the Fe K_α line above background for unknown composition W_1. Next add 3 g of Fe_2O_3 to 97 g of the ore; this will increase the Fe composition

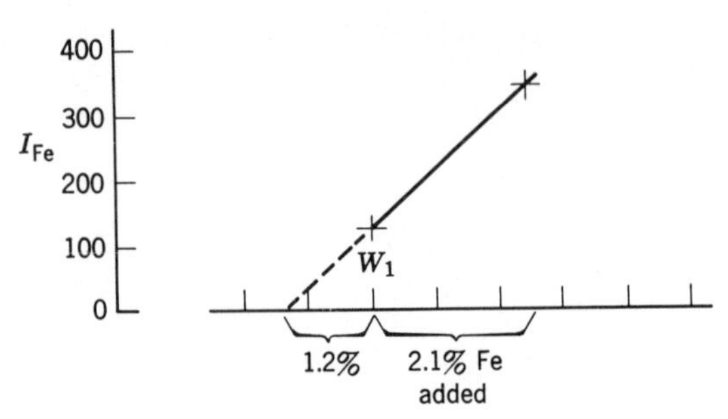

Fig. 7-6. Adding a known amount of the same element as an internal standard.

by $(111.7/159.7) \times 3 = 2.1$ g or 2.1%. This increase is scaled off on the $W\%$ axis of Fig. 7-6 as shown and the intensity of Fe K_α in the new specimen is measured as say 350 cps above background. A straight line may be drawn through the two points and the origin established. Using the same scale divisions for C the numerical value of W_1 is found to be approximately 1.2%. Or we may find W_1 from the equation $W_1/2.1 = 130/(350\text{–}130)$. In practice it is advisable to add two different amounts of Fe_2O_3 to two different aliquots and obtain two additional points instead of only one. A straight line should pass through all three points if the linear approximation is valid; if the points do not follow a straight line within experimental error the best curve may be drawn through the points and extrapolated to the origin.

7.4. Calibration of Solids

For solid samples it is not possible to add internal standards and usually not feasible to prepare a range of compositions in order to plot a calibration curve. Instead, it is commonplace to use comparison standards of known composition similar to the unknown sample. For instance, in analyzing stainless steels comparison standards should be of the same nominal composition as the unknown, i.e., chemically analyzed standards of type 316 steel should be used for analysis of a melt intended to be 316 steel, etc. If this is done the composition for the various elements should agree within 2 or 3% in the unknown and standard and one may use the simple relation

$$W_\text{U}/W_\text{S} = I_\text{U}/I_\text{S}$$

where U and S represent the unknown and standard, respectively, and the I's are the measured intensities above background. This relation corresponds to drawing a straight line from the origin through the point

TABLE 7-2
Nominal Composition for 316 and 410 Stainless Steels

Type	Cr	Ni	Fe	Mo
410	13–15	0	80–83	0
316	16–18	10	68–70	2–3

for the standard and is valid only if the calibration curve does not deviate appreciably from linearity. A slightly more elaborate but better approach is to plot short segments of the calibration curve for each element in the nominal composition range. For instance, in measuring Cr in 316 or 410 steel the nominal composition are shown in Table 7-2. The portions of the Cr calibration curves shown in Fig. 7-7 are drawn by using two chemically analyzed standards of slightly different nominal compositions for each of the steels. These are often called "working curves" and W_U is found by plotting I_U on the calibration segment or from the mathematical relation derived as follows: Let W_1, I_1 and W_2, I_2 be the lower and higher points for the two standards used to establish the appropriate calibration segment. The equations for the straight lines through these points corresponds to $\chi = \mathcal{a}y + b$ where $\mathcal{a} = dx/dy = (W_2 - W_1)/(I_2 - I_1)$ and $b = W_1 - I_1(W_2 - W_1)/(I_2 - I_1)$. Substituting W_U for χ and I_U for y we have

$$W_U = W_1 + I_U(W_2 - W_1)/(I_2 - I_1) - I_1(W_2 - W_1)/(I_2 - I_1) \tag{7-4}$$

W_1, W_2, I_1, I_2 are known for the standards and I_U is the measured intensity above background for the desired element in the unknown sample.

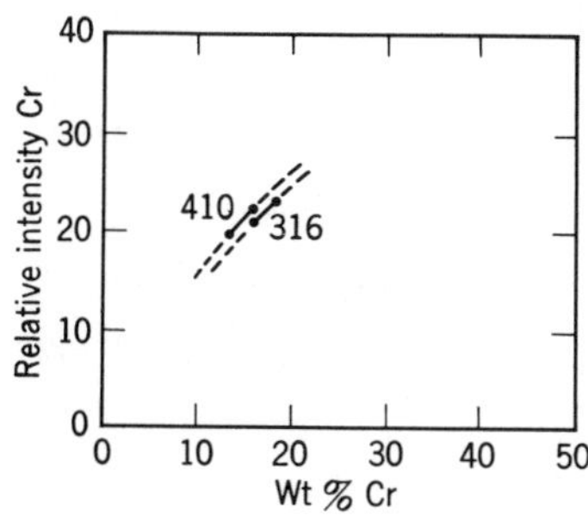

Fig. 7-7. Segments of calibration curves for Cr in types 316 and 410 stainless steels.

7.5. Precision and Accuracy of Analysis

Wet chemical analysis which separates the constituents and weighs them in grams is about the only absolute method of analysis. Accuracy relates the measured composition to the true composition. Other methods of analysis such as X-ray fluorescence are relative methods because the measurable, namely intensity, must be related to composition through calculation or comparison standards. In relative methods of analysis one speaks of precision, which is the agreement between the observed value and the true value of the measurable. A precise value becomes an accurate value only when the conversion to composition is correct. Fortunately, in X-ray fluorescence it is easy to determine precision directly; accuracy is more difficult to establish but comparison with other methods of analysis shows that X-ray accuracy can be made to approach the precision.

X-ray emission obeys the rules for statistics of random processes (radioactive decay, etc.). Therefore the standard deviation, σ, in a measurement of N total counts is just $\sigma = (N)^{1/2}$ and the relative standard deviation, $\sigma\%$, is just $100 \times (N)^{1/2}/N$ or $100/(N)^{1/2}$. For instance, if one sets the crystal spectrometer at the Cr K_α line and measures for 30 sec. to obtain 8450 counts, $\sigma = (8450)^{1/2} = 92$ counts and $\sigma\% = 100 \times 92/8450 = 1.08\%$ which is the precision of the measurement.

X-ray statistics also obeys the usual rule for variance which says that when there are several contributions to error the total variance, σ_T^2, is just the sum of the individual contributions to variance, $\sum \sigma_i^2$. That is:

$$\sigma_T^2 = \sigma_1^2 + \sigma_2^2 + \sigma_3^2 + \cdots \tag{7-5}$$

Since composition is related to the intensity of a line above background it is necessary to consider the variances of the background, σ_B^2, and the total line, σ_L^2, in determining the variance of the line above background, σ_{L-B}^2. If N_L is the total count at the line position and N_B is the total count (for the same counting time) at the background position, then

$$\begin{aligned} \sigma_{L-B}^2 &= \sigma_L^2 + \sigma_B^2 \\ &= N_L + N_B \\ \sigma_{L-B} &= (N_L + N_B)^{1/2} \end{aligned} \tag{7-6}$$

Suppose in the measurement above that the 8450 counts represented N_L and that the background, N_B, for the same counting time was found to be 263 counts. Then

$$\sigma_{L-B} = (N_L + N_B)^{1/2} = (8450 + 263)^{1/2} \approx 93 \text{ counts}$$

and

$$\sigma_{L-B}\% = 100 \times 93/(N_L - N_B) = 9300/8187 = 1.14\%$$

If this measure of 8187 counts above background corresponds to, say, 13% Cr according to the calibration curve then the 1σ limit (67% confidence) is 13% Cr ± 0.15% Cr. Likewise the 2σ limit (95% confidence) is 13% Cr ± 0.30% Cr and the 3σ limit (99% confidence) is 13% Cr ± 0.45% Cr.

If the estimate of composition of an unknown is made from a ratio measurement of an unknown and standard, then the precision of the estimate must include the variance of the standard as well as the unknown. Applying the usual sum for variance, the variance in relative X-ray intensity, R, will be

$$\sigma_R^2 = \sigma_U^2 + \sigma_S^2$$

where σ_U and σ_S are found individually from Eq. 7-6. For instance, if the relative standard deviations (line minus background) for the unknown and standard were found to be 1.3% and 1.4%, respectively, then $\sigma_R\% = (1.3^2 + 1.4^2)^{1/2} = 1.9\%$. Thus it is possible to determine the precision in X-ray analysis rapidly and easily from measurements of the standard and unknown.

It is easy to predict the limit of detectability in X-ray fluorescence by making use of the same statistics. Suppose the limit is *defined* as the value for which the line exceeds the background by 3 standard deviations, that is

$$N_L \geq N_B + 3\sigma_B \tag{7-7}$$

For many specimens the background will be about 3 cps if a suitable high-resolution spectrometer and discrimination against interfering radiation are used. For a 30 sec counting time this will give $N_B = 90$ counts and $\sigma_B = (90)^{1/2} = 9.5$ counts. Thus, $3\sigma_B \approx 28$ counts and, according to the definition N_L must be greater than 118 counts in 30 sec for the element to be detectable.

$$\sigma_{L-B} = (118 + 90)^{1/2} = 14.4$$

$$\sigma_{L-B}\% = 14.4/28 \approx 50\%$$

It is left as an interesting exercise for the reader to show that $\sigma\% \approx 50\%$ no matter what the total N_B is, as long as the definition of Eq. 7-7 is used. One concludes that in X-ray fluorescence analysis the relative standard deviation will always be about 50% at the limit of detectability.

In practice, the limit of detectability ranges from 1 to 10 ppm for most elements[4] because the counting rate approaches 500,000 cps for 100% composition or 0.5 cps for 1 ppm.

7.6. Comparison of Statistics for Fixed-Time and Fixed-Count Measurements

This author does not recommend it, but some analysts prefer to make measurements for lines and backgrounds at a fixed number of total counts for each rather than at a fixed counting time for each. The statistics for fixed counts (FC), is slightly different than for fixed time (FT), but can be compared as follows: Remember that σ_L is the error in the line and σ_B is the error in the background. We will set the ratio of counting rates, $I_B/I_L, = p$.

Fixed Time: For fixed time $N_B/N_L = p$

$$\sigma_{L-B}\%, = 100(\sigma_L^2 + \sigma_B^2)^{1/2}/(N_L - N_B)$$

but

$$\sigma_B^2 = N_B = pN_L = p\sigma_L^2$$

Therefore

$$\sigma_{L-B}\% = 100[N_L(1 + p)]^{1/2}/N_L(1 - p) \tag{7-8}$$

Fixed Count: For fixed counts the counting time, t_B, for the background

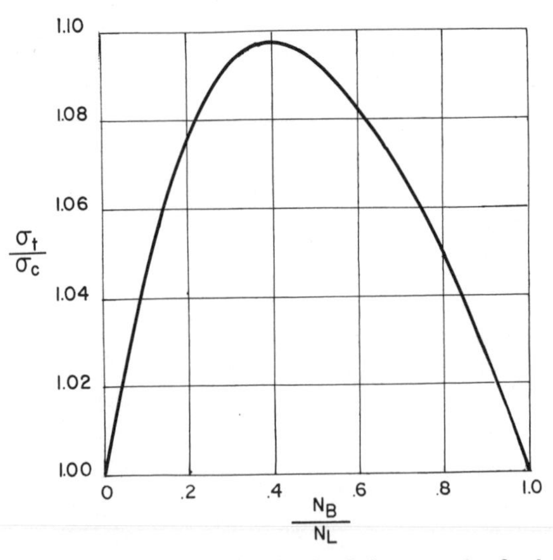

Fig. 7-8. Ratio of standard deviation, σ_t, for fixed time to σ_c for fixed count vs. ratio of background intensity N_B to line intensity N_L.

is extended so that $N_{B(FC)} = N_L = N_{B(FT)}(t_B/t_L)$. This means that the error in the background measurement is reduced by the ratio

$$(N_{B(FT)})^{1/2}/(N_{B(FC)})^{1/2} = (p)^{1/2}$$

That is,

$$\sigma_{B(FC)} = (p)^{1/2}\sigma_{B(FT)}$$

$$\sigma_{L-B}\% = 100(\sigma_L{}^2 + \sigma_B{}^2{}_{(FC)})^{1/2}/(N_L - N_{B(FC)}t_L/t_B)$$

let

$$\sigma_B{}^2{}_{(FC)} = p\sigma^2{}_{B(FT)} = pN_{B(FT)} = p^2N_L$$

and

$$N_{B(FC)}t_L/t_B = N_{B(FT)} = pN_L$$

Therefore

$$\sigma_{L-B}\% = 100[N_L(1 + p^2)]^{1/2}/N_L(1 - p) \qquad (7\text{-}9)$$

Now if we want to know how the error for fixed time compares with that for fixed count we take the ratio of Eq. 7-8 over Eq. 7-9. The denominators, being the same, cancel leaving

$$(FT)\sigma\%/(FC)\sigma\% = [N_L(1 + p)]^{1/2}/[N_L(1 + p^2)]^{1/2}$$
$$= [(1 + p)/(1 + p^2)]^{1/2} \qquad (7\text{-}10)$$

The complete range for p is from $p = 0$ when the background intensity is zero to $p = 1$ when there is no line above the background. Therefore, a plot of the ratio $(FT)\sigma\%/(FC)\sigma\%$ vs. p for $0 < p < 1$ in Fig. 7-8 shows how much worse the uncertainty is for fixed time than for fixed count. In Fig. 7-8 the ratio is unity at p = zero or one and is a maximum at $p = 0.4$. The worst case makes the standard deviation for fixed time only 1.2 times the standard deviation for fixed count and this is usually a trivial difference in precision. Therefore, the fixed-time method is recommended for all measurements because of its simplicity and the shorter time required for background counting.

References

1. W. H. McMaster, N. Kerr Del Grande, J. H. Mallett, N. E. Scofield, R. Cahill, and J. Hubbell, Compilation of X-Ray Cross Sections, Univ. of California, Lawrence Rad. Lab. Report UCRL-50174, 1967, K. F. J. Heinrich, *The Electron Microprobe,* T. D. McKinley, Ed., Wiley, New York, 1966, p. 296.
2. E. L. Gunn, ASTM Special Tech. Pub. 349, p. 70, American Soc. for Testing Materials, Philadelphia, Pa., 1964.
3. F. Claisse, Report P.R. 327, Dept of Mines, Quebec, P.Q., Canada, 1956.
4. W. J. Campbell, J. D. Brown, and J. W. Thatcher, *Anal. Chem. Review Issue,* **38,** 416R (1966).

CHAPTER 8

MATHEMATICAL METHODS FOR QUANTITATIVE ANALYSIS

As explained in Chapter 7 the preparation of calibration curves is always time consuming and frequently difficult. The method of comparison standards requires perhaps a thousand standards for large laboratories doing general analysis, and the analyst may have to try several standards before he finds one that matches the unknown reasonably well. Time is money in analytical procedures and it is worthwhile to reduce the number of standards or calibration curves to a minimum by taking advantage of mathematical aids in evaluating the data. In this chapter we will discuss two mathematical methods which have been found valuable in X-ray analysis. They are called the "Empirical Coefficient Method" and the "Fundamental Parameter Method."[1] Both methods utilize computer programs and give an answer in one or two minutes. The fundamental parameter method is, by far, the most general but the empirical coefficient method is sometimes easier to apply and gives useful results for repetitive analysis of a limited type of specimen.

8.1. Empirical Coefficient Method

The oldest and most common mathematical method uses empirically determined coefficients to represent the matrix effects of one element on another. Even without mathematical proof one recognizes intuitively that the intensity from element i in a matrix must have a unique functional relationship, f, to the composition of all the elements present. The general expression for such a relationship is

$$R_i = f(W_1, W_2, W_3, \ldots) \qquad (8\text{-}1)$$

If the overall matrix effect of element j on element i can be expressed as a constant, α_{ij}, one can write the explicit relation as

$$R_i = W_i / \sum \alpha_{ij} W_j \qquad (8\text{-}2)$$

Here the sum is over all elements including the element i. Eq. 8-2 is called a regression equation.

The validity of Eq. 8-2 depends on two assumptions, both of which are physically incorrect but acceptable if one limits the use of a given set of coefficients to a fairly small composition range in a particular type of specimen such as Fe-base alloys or copper ores or phosphate rock. The assumptions are: (*a*) The polychromatic primary radiation used to excite the specimen can be considered as equivalent to a monochromatic primary radiation of some "effective" wavelength. (*b*) Enhancement can be treated as having the same effect as low absorption.

Although each coefficient $\mathcal{A}_{ij}$ seems to represent just the effect of j on i, it is better to determine $\mathcal{A}_{ij}$ in the same multicomponent system that one wishes to analyze rather than in a sample containing i and j only.* In addition to giving better answers, the determination of $\mathcal{A}_{ij}$ in multicomponent specimens requires the minimum number of initial standards and of the easiest type to obtain. The number of standards is equal to the number of components of interest, that is, if one wishes to analyze for 5 constituents in say phosphate rock, only 5 initial standards are required in order to determine all 25 interelement coefficients. It should be noted that minor constituents can be treated implicitly rather than explicitly as can such elements as oxygen or sulfur, if the elements of interest are present as oxides or sulfides, etc. The method of doing this is explained in Sec. 8.3.

8.2. Determining the $\mathcal{A}_{ij}$ Coefficients

We will use a three-component system of elements A, B, C to show how the coefficients are determined. Three known standards which we will call 1, 2, 3, containing the elements A, B, C are selected and preferably should contain as wide a range of composition as is expected in analysis of the unknown samples. First we measure R_{A1}, R_{A2}, R_{A3}, that is, the relative X-ray intensity from element A in each of the standards. Then we write the simultaneous equations (these are merely Eq. 8-2 with the R and Σ transposed) containing all the known compositions W_{A1}, W_{A2}, W_{A3}, W_{B1}, W_{B2}, etc.

$$\begin{aligned} W_{A1}/R_{A1} &= \mathcal{A}_{AA}W_{A1} + \mathcal{A}_{AB}W_{B1} + \mathcal{A}_{AC}W_{C1} \\ W_{A2}/R_{A2} &= \mathcal{A}_{AA}W_{A2} + \mathcal{A}_{AB}W_{B2} + \mathcal{A}_{AC}W_{C2} \\ W_{A3}/R_{A3} &= \mathcal{A}_{AA}W_{A3} + \mathcal{A}_{AB}W_{B3} + \mathcal{A}_{AC}W_{C3} \end{aligned} \tag{8-3}$$

Solving Eq. 8-3 for $\mathcal{A}_{AA}$, $\mathcal{A}_{AB}$, $\mathcal{A}_{AC}$ may be done by hand for 3 components but is best done by computer for 4 or more components.

* This approach of determining $\mathcal{A}_{ij}$ is multicomponent systems is contradictory to the first edition of this book and to many early papers on the subject. The new approach resulted from a series of experiments conducted at the Naval Research Laboratory in 1964–66 and is largely credited to J. W. Criss.

Next, we measure R_{B1}, R_{B2}, R_{B3} in each of the standards and use them in a similar set of equations to obtain α_{BA}, α_{BB}, α_{BC}. The first equation of that set is

$$W_{B1}/R_{B1} = \alpha_{BA}W_{A1} + \alpha_{BB}W_{B1} + \alpha_{BC}W_{C1}$$

and the others may be written by inspection of Eq. 8-3. Likewise, the coefficients α_{CA}, α_{CB}, α_{CC} are determined by measuring R_{C1}, R_{C2}, R_{C3} and making a third set of simultaneous equations like Eq. 8-3.

8.3. Using the Coefficients in Analysis

Once the 9 coefficients have been determined from only the 3 initial standards we may use them to analyze the concentration W_A, W_B, and W_C in any unknown sample of the same general type used in determining the coefficients. The procedure is as follows: In the unknown sample, measure R_A, R_B, R_C. Write the set of simultaneous equations contain- the 9 coefficients, α_{ij} and the unknown composition to be determined, W_A, W_B, W_C.

$$\begin{aligned} (R_A\alpha_{AA} - 1)W_A + R_A\alpha_{AB}W_B + R_A\alpha_{AC}W_C &= 0 \\ R_B\alpha_{BA}W_A + (R_B\alpha_{BB} - 1)W_B + R_C\alpha_{BC}W_C &= 0 \\ R_C\alpha_{CA}W_A + R_B\alpha_{CB}W_B + (R_C\alpha_{CC} - 1)W_C &= 0 \end{aligned} \tag{8-4}$$

and

$$W_A + W_B + W_C = \text{Constant (see below)} \tag{8-5}$$

These equations are solved for W_A, W_B, and W_C. Equation 8-5 is necessary because the simultaneous equations in Eq. 8-4 are not mutually independent. The constant in Eq. 8-5 is unity if A, B, and C are the only constituents. It may have a value such as 0.98 if the samples have 2% minor constituents which are to be ignored or it may have some other value such as 0.68, 0.83, etc. if A, B, and C are all present as compounds and the compounding element is to be ignored. If the solution of Eqs. 8-4 and 8-5 for W_A, W_B, W_C gives values which do not add up approximately to the constant used in Eq. 8-5, a second estimate of the constant in Eq. 8-5 may be necessary; a value half-way between the original estimate and the sum of W_A, W_B, W_C is probably the best second estimate. Then a new solution for W_A', W_B', W_C' must be found from Eqs. 8-4 and 8-5 and the new values tested again to see if they add up to the estimated constant.

All of the above approach, including the iteration to make the answers consistent with the estimated constant in Eq. 8-5, is easily prepared as a computer program and the complete analysis accomplished in one or two minutes. One precaution should be mentioned. The coefficients α_{ij} were determined for a particular target and voltage and a particular X-ray spectrometer. If a different target material is substituted or if the operating voltage is changed by more than 10% the assumption (*a*) of Sec. 8.1 may no longer ve valid, and it would be necessary to re-determine the coefficients for the new conditions.

8.4. Example of Empirical Coefficients Applied to Analysis of Fe and Ni-base Alloys

The method above was tested with a set of alloys containing the five elements Cr, Fe, Co, Ni, and Mo. The set included 300 series and 400 series stainless steels, an Inconel alloy, and other high-temperature alloys. Each of the specimens had been analyzed by wet chemistry so that all compositions were known and could be compared with X-ray analysis.

The five specimens selected as standards, for calculating the 25 α_{ij} coefficients from equations like Eq. 8-3, were #301, #410, #1187, Hastelloy B, and Inconel X. Table 8-1 shows the values of the α_{ij} coefficients determined. It should be noted that the coefficient values are not the ones which would be obtained if one used binary standards. However they are the best set of 25 values for the matrix effects in the five-component system of alloys; i.e., they take account of all the elements present. The values from Table 8-1 were used with equations like Eqs. 8-4 and 8-5 to analyze the composition of the 6 steels in Table 8-2. The results are shown in the column labeled "X-ray #1." Differences between X-ray analysis and known chemical composition range from a fraction of one percent to several percent and some of the differ-

TABLE 8-1

Empirical Coefficients α_{ij} Used in Test 1 for the Matrix Effect on Element *i* by the Presence of Element *j*

	j				
i	Cr	Fe	Co	Ni	Mo
Cr	1.559	0.516	0.971	0.670	1.731
Fe	2.419	1.023	1.149	0.850	1.286
Co	1.000	1.000	2.184	1.000	1.000
Ni	15.428	−1.110	−7.221	−0.990	5.336
Mo	0.991	1.001	0.110	1.001	1.325

TABLE 8-2
Composition Calculated by Empirical Coefficients

Alloy	Element	Chemical (%)	X-ray #1 (%)	Difference (%)	X-ray #2 (%)	Difference (%)
303	Cr	17.2	17.8	0.6		
	Fe	71.2	73.8	2.6	Used as standard	
	Ni	8.7	8.4	−0.3		
304	Cr	18.6	18.8	0.2	19.0	0.4
	Fe	69.5	71.0	1.5	71.7	2.2
	Ni	9.4	9.7	0.3	9.3	−0.1
316	Cr	17.7	17.8	0.1		
	Fe	64.8	66.0	1.2	Used as standard	
	Ni	12.8	13.2	0.4		
321	Cr	17.8	18.5	0.7	18.4	0.6
	Fe	68.2	71.7	3.5	70.6	2.4
	Ni	10.8	10.9	0.1	11.0	0.2
347	Cr	17.7	18.5	0.8	18.4	0.7
	Fe	67.9	71.7	3.8	70.2	2.3
	Ni	10.7	11.7	1.0	11.4	0.7
430	Cr	17.5	17.6	0.1	17.5	0.0
	Fe	81.3	79.3	−2.0	81.4	0.1

ences may represent errors in measurement as well as limitations of the method.

Next, a new set of 25 $\mathcal{a}_{ij}$ coefficients was determined using only 300 series steels as the initial 5 standards and the analyses repeated with these coefficients. The results are shown in Table 8-2 in the column labeled "X-ray #2." Again the differences are about the same as before indicating that the large differences may well have been due to experimental errors. None of the coefficients obtained for this system of elements would be applicable to the analysis of these elements in ores such as chromite or olivine because the remainder of the matrix is completely different.

8.5. Fundamental Parameter Method

In contrast to the empirical coefficient method, the fundamental parameter method assumes only that the specimen is homogeneous, thick, and has a reasonably flat surface. Instead of assuming that the incident spectrum can be described by a single average wavelength, it uses the measured primary spectral distribution for a given target and operating voltage. The matrix absorption and secondary fluorescence enter explicitly for each specimen, and composition is calculated directly by iteration. The advantage is that no intermediate standards or empirical coefficients are needed for any matrix. The disadvantage is the present uncertainty in the mass absorption and fluorescent yield parameters. This shortcoming is being overcome rapidly, however, as more laboratories devote themselves to remeasuring the parameters accurately.

The first parameter needed is the spectral distribution of the primary radiation. This depends on the tube-target material, the voltage, and the tube-window thickness. Tabulation of spectral distributions for a variety of targets and voltages are given in Appendix 1. Figure 3-2 showed the measured spectrum for a tungsten target at 50 keV fullwave rectified operation and a 0.030 in. Be window. The discontinuity at the L_{III} edge (1.22 Å) is due to the change in target absorption for emerging radiation at that wavelength. For each wavelength interval, $\Delta\lambda$, in the primary spectrum the integrated intensity, $I_\lambda \Delta\lambda$, is measured from the curve. The intensity of the characteristic tungsten L lines is, of course, included at the appropriate wavelength intervals. This integrated value, $I\lambda\Delta\lambda$, is the parameter for the primary intensity at each wavelength.

8.6. Computer Program and Equations

A computer program is set up to calculate (*a*) the penetration of each incident primary wavelength, λ, into the specimen according to Fig. 8-1, (*b*) the primary excitation of each element in layer dx, (*c*) the secondary fluorescence of each element in each layer dy due to characteristic radiation from layer dx, and (*d*) the absorption of each emerging characteristic radiation. It is necessary to assume some composition to start the

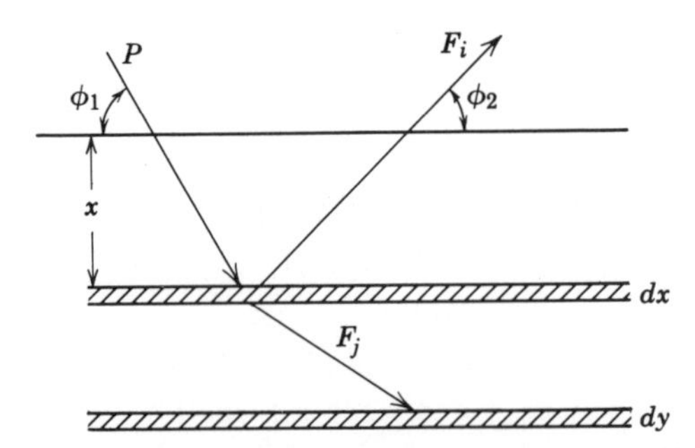

Fig. 8-1. Geometry for calculation of primary and secondary contributions to characteristic intensity by fundamental parameter method.

computer calculation; the simplest assumption is just the measured relative X-ray intensities for each element scaled so that the sum equals 100% (or some other total if minor constituents are to be ignored). The computer then calculates what intensities should have been observed for the assumed composition, compares them with the measured values, adjusts the assumed composition accordingly, and calculates a new set of expected intensities. This iteration process is repeated automatically until the assumed composition gives expected X-ray intensities which match the measured values within some prescribed agreement (usually 0.1%). The composition for which this occurs is printed out as the analysis.

The equations used in the computation employ standard tables of mass absorption coefficients and fluorescent yields. The equations are shown below only to indicate that, although lengthy, they are not complex mathematically. Complete solution with 3 or 4 iterations takes less than 1 min on a CDC 3800 computer and costs less than $0.50 per element. Time and costs on other computers depend on the specific computer used.

The contribution to the intensity from element i in matrix M due to primary excitation only is called P_{Mi}; the corresponding intensity from the pure-element standard is P_{ii}. The contribution due to sceondary fluorescence in the matrix is S_{Mi} and there is no secondary fluorescence in the pure-element standard (if some intermediate composition is used as the normalizing standard a contribution due to secondary fluorescence would have to be considered). The relative X-ray intensity, R_i, is given by

$$R_i = (P_{Mi} + S_{Mi})/P_{ii} \tag{8-6}$$

where

$$P_{Mi} = g_i W_i \sum \frac{D_{i\lambda}\mu_{i\lambda}I_\lambda \Delta\lambda}{\mu_{M\lambda} \csc \phi_1 + \mu_{Mi} \csc \phi_2} \tag{8-7}$$

P_{ii} is a similar equation except that $W_i = 1$ and $\mu_{i\lambda}$ and μ_{ii} are substituted for $\mu_{M\lambda}$ and μ_{Mi} in the denominator. The secondary fluorescence contribution S_{Mi} is given by

$$S_{Mi} = g_i W_i \sum \left\{ \frac{1}{\mu_{i\lambda}} \sum_j \left(D_{j\lambda} W_j K_j \mu_{ij} \mu_{j\lambda} \right.\right.$$
$$\times \left[\frac{1}{\mu_{M\lambda} \csc \phi_1} \ln \left(1 + \frac{\mu_{M\lambda} \csc \phi_1}{\mu_{Mj}} \right) \right.$$
$$\left.\left.\left. + \frac{1}{\mu_{Mi} \csc \phi_2} \ln \left(1 + \frac{\mu_{Mi} \csc \phi_2}{\mu_{Mj}} \right) \right] \right) \right\} \tag{8-8}$$

In Eqs. 8-7 and 8-8 the term g_i is related to absolute intensity but cancels since it appears in both the numerator and denominator of Eq. 8-6. $D_{i\lambda}$ is merely a constant of zero or unity depending on whether or not the particular primary wavelength λ has enough energy to excite element i; $D_{j\lambda}$ is a similar constant for element j. The first subscript for μ is the absorber (matrix M, or element i or element j); the second subscript refers to the radiation being absorbed (primary wavelength λ or characteristic radiation of element i or element j). ϕ_1 and ϕ_2 are the incident and emergent angles in Fig. 8-1. $K_j = [1 - (1/J)]\ \omega_j$ where ω_j is the fluorescent yield (Appendix 3) for the particular line of element j which will excite element i; J is the absorption edge jump factor, i.e., the ratio of absorption coefficients on the two sides of the absorption edge; J has a value of about 10 for the K edges and is tabulated in Ref. 2.

TABLE 8-3
Composition Calculated by Fundamental Parameters

Type	Element	Chemical (%)	X-ray (%)	Difference(%)
303	Cr	17.2	18.5	1.3
	Mn	1.3	1.3	0.0
	Fe	71.2	71.5	0.3
	Ni	8.7	8.7	0.0
304	Cr	18.6	19.8	1.2
	Mn	1.4	1.5	0.1
	Fe	69.5	69.6	0.1
	Ni	9.4	9.1	−0.3
316	Cr	17.7	19.0	1.3
	Mn	1.8	1.8	0.0
	Fe	64.8	66.5	1.7
	Ni	12.8	12.7	−0.1
321	Cr	17.8	19.0	1.2
	Mn	1.6	1.7	0.1
	Fe	68.2	68.7	0.5
	Ni	10.8	10.6	−0.2
347	Cr	17.7	18.9	1.2
	Mn	1.6	1.7	0.1
	Fe	67.9	68.5	0.6
	Ni	10.7	10.9	0.2
430	Cr	17.5	18.8	1.3
	Fe	81.3	79.6	−1.7

8.7. Example of Fundamental Parameters Applied to Analysis of Fe and Ni-base Alloys

The computer program of Sec. 8.6 was used to analyze the same Fe and Ni-base alloys used in Sec. 8.4 but it was not necessary to use any of the alloys as standards. Table 8-3 shows the results. Mn was included in these calculations because any element or number of elements can be considered; Mn could not be included in Table 8-2 because no empirical coefficients had been determined for Mn. The agreement between chemical and X-ray values by the fundamental parameter method in Table 8-3 is slightly better than it was by the empirical coefficient method of Table 8-2.

8.8. Future Directions for Calculation Methods

As access to high-speed computers becomes commonplace, the fundamental parameter method or even more complex methods such as three-dimensional Monte Carlo calculations will be used routinely because of their general applicability to all types of specimens. The primary spectral distribution, once measured for a given target and voltage can be used by all laboratories as can the tables for μ and ω. By contrast, each laboratory must prepare its own empirical coefficients for its particular spectrometer and for each general type of specimen to be analyzed.

With present computers it is the input and output time which is costly; thus more elaborate calculations inside the computer add little to the overall cost.

References

1. J. W. Criss and L. S. Birks, *Anal. Chem.*, **40**, 1080 (1968).
2. T. D. McKinley, K. F. J. Heinrich, and D. B. Wittry, Eds., *The Electron Microprobe*, Wiley, New York, 1966, p. 296.

CHAPTER 9

APPLICATIONS AND SPECIMEN PREPARATION

Applications can be categorized in several ways such as shown in Table 9-1. Specimen preparation, equipment required, and accuracy of analysis depends more on the last three columns in the table than on the field of interest. For instance, in a solid sample where one is interested in a low concentration of sulfur the problems would be somewhat similar whether the sample be a steel alloy, a mineral, a fired ceramic, or a biological section. That is, one would have to prepare a smooth surface, use a long-spacing crystal and a thin-window detector, and measure for sufficient time to obtain suitable statistics for the line above background. The general types of specimen preparation include cutting, grinding, polishing, chemical solution, and concentration. In the sections to follow, examples are discussed for some of the categories but the analyst may have to improvise for his specific problem.

9.1. Solids

Some solids such as fired ceramics, glass, or porcelain enamel are smooth enough so that no surface preparation is required. Other hard samples such as alloys or minerals must be ground flat using carborundum, sapphire, or diamond abrasives. The grinding marks, Fig. 9-1, should be oriented parallel to the plane of incident and emerging X-rays to insure minimum variation in absorption effects due to the furrows. Soft materials such as biologicals or plastics may be cut with a microtome knife or similar device but care must be used to avoid smearing. In order to test for the minimum specimen preparation required it is wise to compare X-ray intensities for the same sample in several stages of smoothness. For instance, in steel alloys one might examine the specimen after each graded grinding or series of abrasive papers and after metallurgical polishing on a cloth wheel. The X-ray intensities of the various elements will reach constant values after some stage of grinding and hold these values through the fine polishing. Routine preparation can be stopped with confidence at the coarsest stage of grinding that was shown to yield the same intensities as the fine polishing.

It should be noted that the smoothness requirements depend on the atomic number of the elements to be analyzed. That is, a rougher

TABLE 9-1

Applications Categorized in Various Ways
(The columns are unrelated to each other.)

By field of interest	By physical state	By concentration range	By atomic number
alloys	solids	1–100%	>22
minerals	powders	0.01–1%	12–22
ores	liquids	1–100 ppm	5–12
petroleum	gas	<1 ppm	
glass	heterogeneous		
ceramics			
pigments			
biologicals			
plastics			
thin films			

surface can be tolerated if one is measuring Ni K_α in steel than if he is measuring S K_α or Si K_α in the same matrix. This is because the Ni K_α will be less strongly absorbed than S K_α or Si K_α and therefore surface roughness will have less effect. As a rule-of-thumb, the surface roughness should be less than the path length for 10% absorption of the radiation of interest. For instance, in measuring S K_α in steel the matrix absorption for type 316 steel is approximately

$$\begin{aligned} \mu_{MS} &= \mu_{FeS}W_{Fe} + \mu_{NiS}W_{Ni} + \mu_{CrS}W_{Cr} \\ &= 1150 \times 0.7 + 1200 \times 0.1 + 920 \times 0.18 \\ &= 1091 \end{aligned}$$

The path length, t, for 10% absorption (90% transmission) of S K_α is

$$\begin{aligned} I/I_0 = 0.9 &= \exp(-\mu_{MS}\rho t) \\ 0.90 &= \exp(-1091 \times 7.9 \times t) \end{aligned}$$

$$t = 0.104/\mu_{MS}\rho = 1.21 \times 10^{-5}\ \text{cm} = 0.12\ \mu \approx$$
$$6 \text{ millionths of an inch}$$

For solid samples, comparison standards should be similar in composition and with the same surface preparation as the unknowns. Families of calibration curves will be required as discussed in Chapter 7. The mathematical calculations of Chapter 8 are usually suitable for solids unless they be heterogeneous.

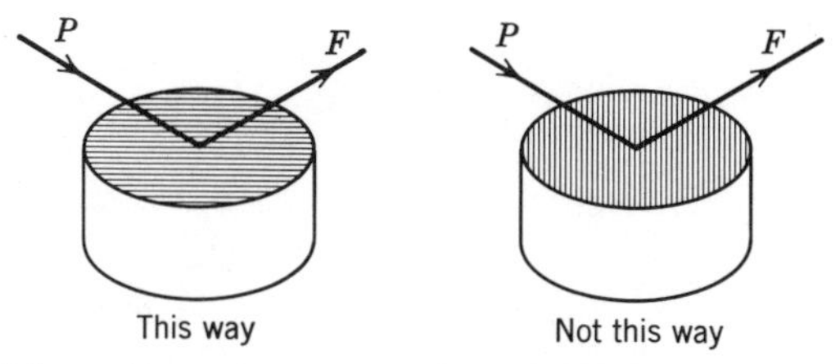

Fig. 9-1. The right and wrong ways to orient abraded surface with respect to the directions of primary, P, and fluorescent, F, radiation.

9.2. Powders

Powder samples fall into several types. Paint pigments are usually homogeneous or mixtures of homogeneous particles. Ores may contain mixtures of homogeneous particles or each particle may be cored, layered, or otherwise heterogeneous. Cements and phosphate rocks may be mixtures of different compounds. One generally satisfactory preparation is to grind these powders fine enough to satisfy the 10% absorption rule of Sec. 9.1 for the elements to be measured. A mesh size of -300 has been found satisfactory for many applications, but, again the analyst should test by examining his particular specimen type in several stages of grinding in order to reach constant X-ray intensities for the various elements of interest.

Dry powders can be packed in cells or pressed into pellets with or without binder. Powders can also be examined in slurry form in the case of ore samples. The amount of binder in the case of dry powders or liquid in the case of slurries should be the same for unknowns and standards and the preparation should always be as similar as possible. Calibration curves are somewhat easier to prepare for powders than for solids because additions are readily made.

Some samples such as Cu and Fe ore slurries pose special problems because the metal elements are present as mixed oxide and sulfide compounds and the matrix absorption for Cu K_α and Fe K_α depends on the relative amounts of oxygen and sulfur. In such cases it is often easier to measure the matrix absorption coefficient directly and use this as an adjustment on the Cu and Fe intensities rather than to measure the S K_α or O K_α intensities. Matrix absorption may be measured by transmission of the incident radiation through the cell containing the slurry but is better done by introducing a metal plate in the bottom of the cell and measuring its characteristic fluorescence intensity. Figure 9-2 shows such an arrangement. The plate element should be one whose characteristic radiation is absorbed differently by the oxygen and the sulfur.

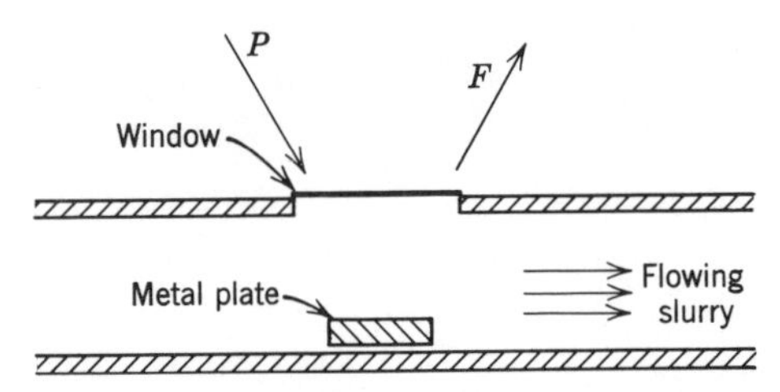

Fig. 9-2. Use of a fluorescing metal plate in a flow cell to determine matrix absorption of the slurry.

9.3. Liquids

Liquids are regarded by some analysts as the ideal sample because they are certain to be homogeneous and because comparison standards and calibration curves are most readily prepared. Specimen preparation requires no discussion but some of the problems attendant to examination need to be considered.

Some natural liquids such as gasoline are volatile and must be contained in a closed cell. The cell window must, of course, be impervious to the liquid and should be thin enough to transmit both the incident and fluorescent radiation without appreciable attenuation. Often the interaction of the X-rays with the liquid may cause bubble formation and if these bubbles collect on the window the X-ray intensities will be affected. For such liquids an inverted cell as shown in Fig. 9-3 is advantageous because the bubbles are less likely to collect on the window.

Most liquids will have lower X-ray absorption than solids or powders and therefore the cells should be deep enough to approximate an infinitely thick layer. This is not always convenient or even possible, however, and one is forced to work with cells of lesser depth. In such cases the cells used must be uniform in thickness to 1–5% to insure reproducibility of data. The back wall of the cells must also be uniform

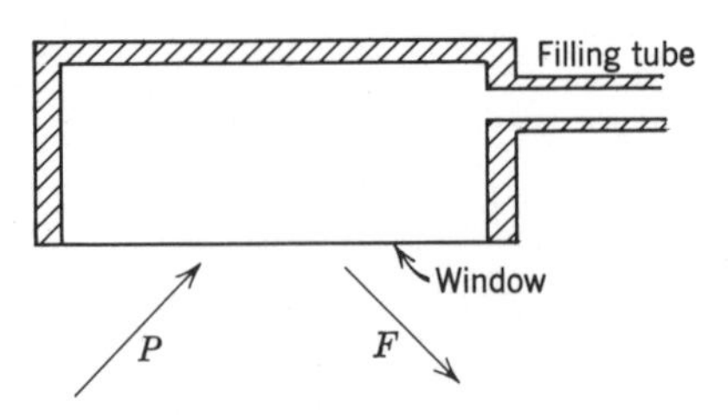

Fig. 9-3. Inverted cell arrangement with thin window to contain the liquid.

in thickness to insure a constant back-scattering of primary radiation. An internal standard is often added to a liquid cell by placing a metal plate in the bottom of the cell much the same as the plate in Fig. 9-2.

Nonvolatile liquids can be examined in upright cells provided that the liquid level is maintained at a uniform value. Again, the cell depth must be carefully controlled if the sample is of less than effectively infinite thickness. Liquids lend themselves readily to "on-stream" analysis because they can be flowed continuously through a cell. This is discussed more fully in Sec. 9.9. Liquids may be transformed to solids by freezing or by forming a gel or grease. For instance oils may be mixed with lithium stearate to form a grease and the stearate does not absorb the radiation appreciably. In some liquid samples precipitation may occur due to interaction with the X-rays. This will change the X-ray intensities drastically and such samples are not suitable for quantitative analysis.

9.4. Solid Solutions (Borax Bead)

A useful technique for solids or powders which are heterogeneous or which suffer from strong matrix effects is to dissolve the sample in melted borax or other similar material.[1] The borax is melted in a platinum crucible and 1–5 parts of sample are added to 10 parts of borax. The mixture should be stirred to achieve homogeneity. Then the melt is cast onto a cool plate which forms a flat surface on the bottom of the bead and is suitable for examination without further preparation. Sometimes the beads crack on cooling and improvisation or variation of concentrations or temperature may be necessary for specific samples. The borax beads are stable with time and may be stored conveniently for future reference. As with all dilution techniques, the X-ray intensities are reduced but this may well be preferable to heterogeneous samples.

9.5. Thin Films and Plating Layers

Whenever a sample is not thick enough to absorb all of the incident primary radiation the X-ray intensities will depend on the absolute amount of each element as well as the composition. There are two types of specimens that generally fall into this category. One type includes evaporated or plated metal layers. The other type includes paint layers, oxide films stripped off metals, corrosion layers, etc.

Evaporated or plated layers are the simplest type because one knows the element and is only interested in the thickness. The layer is usually

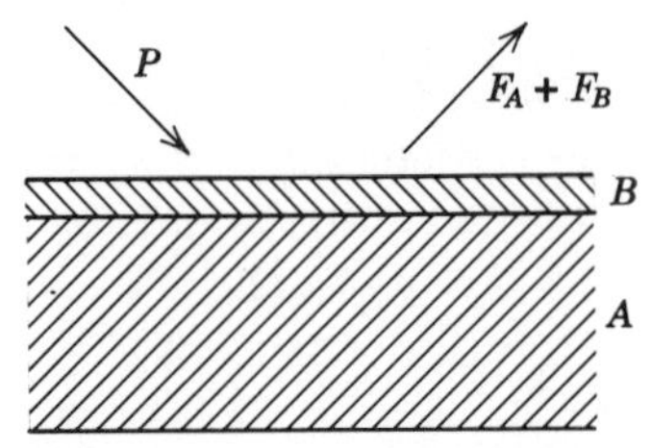

Fig. 9-4. Measurement of plating layer thickness using either fluorescence of the plating layer, B, or of the base material, A.

examined directly on the substrate, Fig. 9-4, and the characteristic X-rays of either the plating or the substrate may be used depending on the particular elements involved. Tin-plate on steel was one of the first applications of X-ray fluorescence[2] and is still a standard technique. The Fe K_α intensity is strongly absorbed by Sn so that the method is very sensitive to small changes in Sn thickness. No spectrometer is necessary for tin-plate measurements because the X-ray tube can be operated at less than 25 keV so that only Fe K_α radiation is excited. For Cr plate on steel it would also be advisable to measure the absorbed Fe K_α intensity because Cr is also present in the steel itself. On the other hand, for Ni plate on Cu either the emitted Ni K_α or the strongly absorbed Cu K_β could be used as a sensitive measure of the Ni thickness. Calibration curves can be prepared from known thicknesses or, if the substrate intensity is used, the absorption equations (Eq. 2-4b) can be used for a direct calculation if the density of the plating layer is known or assumed.

Multiple plating layers often occur in practice and the thickness of each layer can be determined by a combination of measurements and a family of calibration curves. Usually the outermost layer will be determined by measuring its characteristic intensity and the next layer will be determined by measuring its characteristic intensity and correcting for absorption by the now known outer-layer thickness.

Considering the various methods and combination of methods, plating layers from about 1 μm to perhaps 0.5 mm may be determined. Precision ranges from about 50% (the limit of detectability, see Sec. 7.5) for the thinnest layers to 1% for intermediate layers around 50 μm (0.002 in.) and back to 50% for the thickest layers around 0.5 mm. The plating material and the substrate will cause variation in both the thickness measurable and the precision.

Thin films of unknown composition can be analyzed by measuring the relative X-ray intensities provided that the films are thin enough so that matrix absorption and fluorescence can be neglected. Less than 5% total absorption for any of the characteristic rediations of interest might be considered negligible. As an example, suppose one has Fe–Ni films with as much as 50% Fe. The thickness limit will be set by Ni K_α which is the most strongly absorbed.

$$\mu_{\mathrm{MNi}} = \mu_{\mathrm{FeNi}} W_{\mathrm{Fe}} + \mu_{\mathrm{NiNi}} W_{\mathrm{Ni}}$$
$$= 400 \times 0.5 + 60 \times 0.5 = 230$$

For 95% transmission

$$I/I_0 = 0.95 = \exp(-230\,\rho x)$$
$$\rho x = 0.0487/230 \approx 200\ \mu\mathrm{g/cm}^2$$

Usually the film must be supported on a thin backing layer such as Mylar, and a beam stop should be used behind the sample to prevent scattering of the transmitted primary radiation. If intensity for each element is to be calibrated in terms of μg/cm^2 or similar units, standards of known thickness must be prepared for each individual element. Percent composition is obtained by dividing the μg/cm^2 for an element by the total number of μg/cm^2 for all elements. Using thin oxide layers stripped from stainless steel, Rhodin[3] was able to measure microgram quantities of iron, nickel, and chromium by this technique; his results are given in Table 9-2.

TABLE 9-2
X-Ray Intensities for Microgram Quantities

Element	Intensity[a]
Cr	8.5
Fe	16.5
Ni	26.8

[a] Numbers are in (counts per second)/(micrograms per square cm). (After Rhodin, Ref. 3.)

9.6. Low Concentrations

Concentrations as low as 1–10 ppm may be measured for many elements. Table 9-3 shows observed limits detected by Campbell[4] using the criterion of Sec. 7.6. The values depend on the matrix, of course, and are generally lowest for metal elements in biological materials where matrix absorption is low. The counting rate is usually a few counts per second per microgram in favorable situations; counting times of up to

TABLE 9-3

Detectability Limits[a]

Element	ppm	Element	ppm
A. In Iron Matrix (10 min counting time)			
Si K_α	170		
Ti K_α	1.0	Cu K_α	8.5
V K_α	1.9	Ar K_α	6.8
Cr K_α	4.0	Zr K_α	4.6
Mn K_α	1.4	Mo K_α	4.5
Ni K_α	5.4	Sn L_α	3.9
B. In Water Matrix (2 min counting time)			
S K_α	140	Cd L_{β_1}	10
Cl K_α	62	I K_α	12
K K_α	3.2	I L_α	4.2
Ca K_α	2.3	Ba K_α	14
V K_α	1.5	Ba L_α	4.8
Cr K_α	5.1	La K_α	20
Fe K_α	1.8	La L_α	5.3
Cu K_α	2.4	Sm K_α	4.1
Ar K_α	1.8	Sm L_α	10
Sr K_α	1.5	Yb L_α	6.8
Mo K_α	2.0	Au L_α	19
Mo L_{β_1}	49	Pb L_α	5.5
Cd K_α	5.4	Th L_α	6.5

[a] After Campbell, Ref. 4.

5 min are necessary for reasonable statistics. The limits for low Z elements occur at higher concentrations because of the low fluorescent yield, lower diffraction efficiency of crystals, and window absorption in X-ray tubes and detectors.

Preconcentration is valuable for low concentrations. It may take the form of chemical solution and either precipitation of the desired elements or use of ion exchange membranes to entrap the desired elements. Precipitates are often collected on filter paper, dried, and the filter placed directly in the X-ray apparatus. Likewise, ion-exchange-resin-loaded papers[5] make a suitable sample in the X-ray equipment. Theoretically there are no limits on the concentrations detectable if one is willing to start with a large enough sample for solution. Practically, however, the limit for X-ray fluorescence is somewhere in the neighborhood of 0.01–0.1 ppm because other analytical techniques such as neutron activation may be less costly than the extensive preparation required for X-ray fluorescence analysis of lesser amounts.

9.7. Low Atomic Number Elements

For atomic numbers below about 22 (Ti) fluorescent yield is low and the longer wavelengths are absorbed strongly by the air paths in the spectrometer and in the X-ray tube and detector windows. The lower atomic number elements divide themselves into two ranges, namely Na to Ti and B to Na because of the type of equipment required.

From Na to Ti a vacuum or helium path may be used in the spectrometer. Thin window Cr or Ag target X-ray tubes can be used although pumped tubes are advantageous for Na to Si. Flow proportional counters are necessary; ¼-mil Mylar windows and 90% argon–10% methane gas are used in these detectors. KCl crystals can be used down to S K_α (5.4 Å) but KAP, EDDT, or pentaerythritol (see Appendix 2) are required in order to go down to Na. A word of caution about KAP: the potassium is strongly excited by the radiation elements from Cu, Ti, etc. and may give unacceptably high background if such elements are present in the sample.

For elements from B to Na direct electron excitation or very thin window, continuously pumped X-ray tubes are required. Both the tube and detector windows should be of material such as cast Fromvar,[6] nitrocellulose, or stretched polypropylene and be less than 1 μm thick to avoid excessive attenuation. The detector gas is usually a He–methane mixture or just methane. A vacuum spectrometer is required. KAP crystals can conveniently be used down to fluorine but multilayer films or organic crystals such as OHM[7] (see Appendix 2) are required for B, C, N, O. The multilayer films such as barium stearate or lead stearate decanoate[6] are prepared by casting a monolayer of the material on water and dipping a glass slide in and out through the monolayer to form 50–200 layers which will diffract the long wavelength X-rays. Organics such as OHM are quite expensive but may replace the stearate layers if they can be grown in large numbers and with sufficient reliability. With suitable equipment, concentrations as low as 0.1% C in steel can be measured directly.

9.8. Heterogeneous Samples and Limited Quantities

For most X-ray fluorescence analysis one is interested in the average composition of more-or-less homogeneous samples. When the sample is heterogeneous and one wishes to measure the variation in composition or when one has a very limited quantity (a few mg or less) of material, it is necessary to mask the primary X-ray beam on the spectrometer so that a localized area is measured. Several possibilities exist and have been

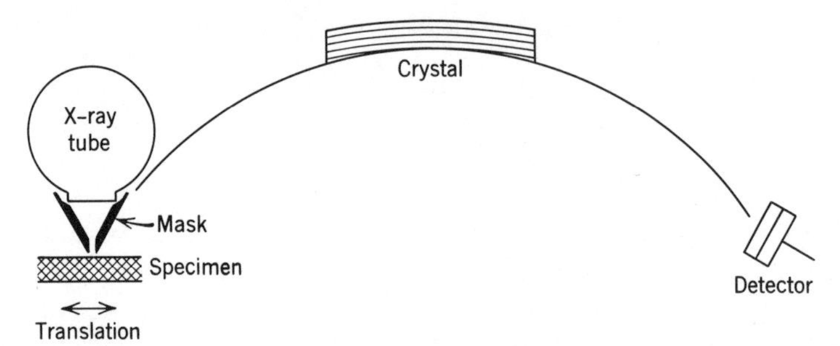

Fig. 9-5. Measuring variation in composition across a specimen using a masked primary beam and a curved-crystal spectrometer.

applied in practice. The greatest intensity from a masked specimen is achieved with the curved-crystal spectrometer arrangement shown in Fig. 9-5. A lead mask limits the primary radiation to a circle of 0.2–0.5 mm on the specimen which can be translated to bring the desired area into the beam. By scanning the specimen the variation in composition across the surface can be plotted in the form of a topographic map. The same arrangement can be used with quantities of material as small as micrograms or less if the material can be collected in the small area irradiated. For quantities as large as 1 mg the intensity compares with that from a full-sized specimen in a flat-crystal spectrometer, Table 9-4.

TABLE 9-4
X-Ray Intensity from an Aluminum Specimen Containing 0.52% Fe and 0.81% Mn

Element	Flat crystal optics extended specimen	Curved crystal optics 1 mg specimen
Fe	280 cps	360 cps
Mn	190	160

The masking arrangement is also used with standard flat-crystal spectrometers at the expense of intensity. Either the primary beam is masked so that a limited area of the specimen is irradiated or a mask is used to limit the area of the specimen which can be seen by the spectrometer. Such an arrangement is often referred to as a milliprobe[8] in contrast to the electron microprobe discussed in Chapter 10.

9.9. On-Stream Analysis and Process Control

Because of the dynamic nature of X-ray intensity measurements, the method lends itself the most readily of all analytical techniques to continuous process control. This is used widely in on-stream analysis of or slurries, gasoline additives, cement powder production, etc. The objective is to flow the sample past the X-ray equipment and measure the variation in several elements simultaneously with several detectors and associated electronics. Either an open loop where the data are merely recorded or a closed loop where the results are used to control the process may be employed.

In the case of ore slurries the flow cell may take the form of Fig. 9-6 or the previous Fig. 9-2. The flow rate and design of the cell must not allow pile-up of solids at corners. Solids concentration must be determined independently by absorption measurements or by fluorescence from a metal plate in the cell. Standards may be introduced occasionally to calibrate the equipment and mathematical analysis of the data by one of the methods of Chapter 8 is advantageous because of its continuous nature and speed.

Addition of tetraethyllead to gasoline lends itself readily to closed-loop continuous process control because any variation in the Pb L_α line intensity can be used immediately to adjust the addition. This is a somewhat different and easier problem than the measurement of varying

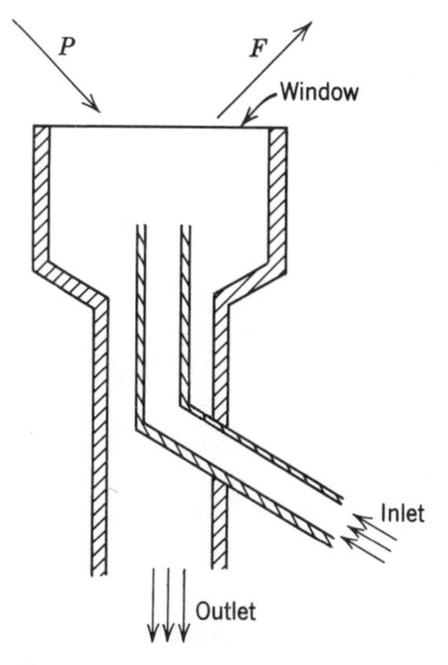

Fig. 9-6. End-window flow cell for on-stream analysis.

concentration of metals in ore slurries. Other chemical production problems are likely to be similar to the gasoline problem.

In cement powder production part of the continuously moving dry material is compressed into pellets and the pellets read individually but automatically.[9] Again the objective is constant composition. Any variation outside the set limits is used to adjust the process but usually not in a closed-loop fashion.

Production control is being used in the steel industry in a slightly different manner. As the melt approaches the time for pouring, last-minute additions are necessary in order to achieve the desired analysis. A rapid sequence of samples is used to follow the changing composition of the melt. The mathematical methods of Chapter 8 are necessary for the rapid computations required and the results are used with open-loop or closed-loop to control the additions.

9.10. Valence Effects on X-Ray Emission Lines

In Chapter 2 it was stated that one of the advantages of X-ray fluorescence analysis was that the characteristic lines were nearly independent of the physical state or chemical combination of the elements. This is less true for K lines of low Z elements and L or M lines of higher Z elements where there is some shifting in wavelength and distortion of line shape as the valence state of the atom changes. Figure 9-7 shows how the C K_α line is affected by valence state. Although this initially appears as a difficulty if one wished to measure only carbon composition, it can be turned to an advantage if one also wishes to know

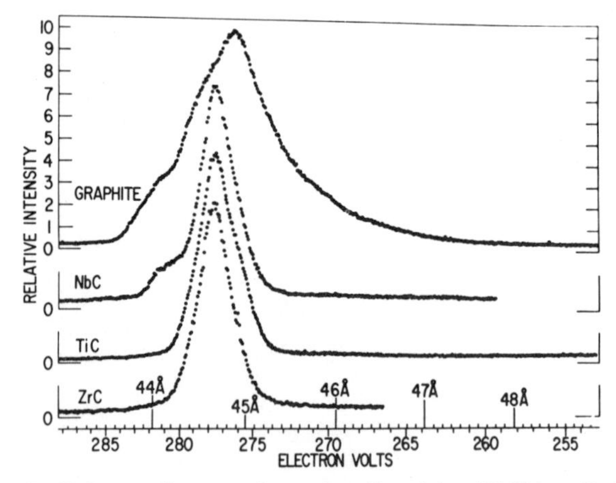

Fig. 9-7. Valence effects on the carbon line (after Holliday, Ref. 10).

how the carbon is combined chemically. For elements such as sulfur where many valence states are possible and may occur in a single sample, two alternative approaches suggest themselves: (*a*) Sulfur line shapes from individual standards of known-valence may be used to unfold the separate contributions to a complex S K_α line from an unknown sample. (*b*) Spectrometer resolution may be degraded deliberately to smear the line out to uniform width with the hope that peak height can be related to total sulfur content. Computer programs will be of considerable value in treating either situation for quantitative analysis. Much work remains to be done in elucidating the effects of valence and in employing them in chemical analysis.

An important application of X-ray emission line shape in the physics of solids is also developing from recent work. When atoms share electrons as they do in alloys, the electron levels are changed slightly. This changes both the absorption fine structure and the emission line shapes. Figure 9-8 shows the Al K_α', K_{α_3}, K_{α_4} lines in Fe–Al alloys.[11] As the amount of Fe increases the average number of electrons in the $3p$ shell of Al decreases, and the line shifts to shorter wavelengths. By measuring

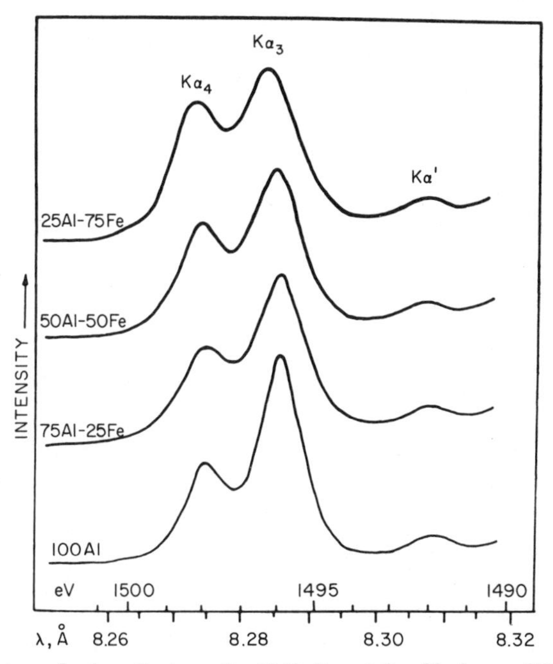

Fig. 9-8. Electron sharing effects on the Al K_α lines (after Fischer and Baun, Ref.11).

such shifts in the L or M lines of higher Z elements it should be possible to tell which atom is the donor of electrons and which the receiver in alloys. In other atoms such as the rare earths where the $4f$ shell is unfilled, or in the atoms around Fe where the $3d$ shell is unfilled, ground state atoms may absorb the characteristic photons emitted by excited atoms and thus distort the emission line shapes. Figure 9-9 shows the situation for the rare earths and Fig. 9-10 shows how the M_α line from Tb to Lu is affected by the changing number of $4f$ electrons in this self-absorption process.[12] Such applications fall more properly under the heading of X-ray spectroscopy rather than X-ray spectrochemical analysis but the same type of equipment is used.

9.11. Indirect Analysis

Because of the difficulties in measuring the characteristic lines of low Z elements and because it is not feasible to measure the lines below B, indirect methods have sometimes been used for analysis. Dwiggens[13] used Compton scattering to measure the H/C ratio in hydrocarbons. Compton scattering increases with decreasing atomic number and is considerably greater for H than for C. The intensity of the Compton scattering of the W L_α line of the primary radiation by hydrocarbons is therefore dependent on the H/C ratios and may be used to measure it. Table 9-5 shows Dwiggens' results.

Another indirect method is to dissolve the sample and precipitate the desired element in a compound where the characteristic lines of the added element are more easily measured. Smith[14] precipitated Si in a uranium compound and used U L_α as a measure of the Si content of the original sample; he also precipitated K as a silver–cobalt com-

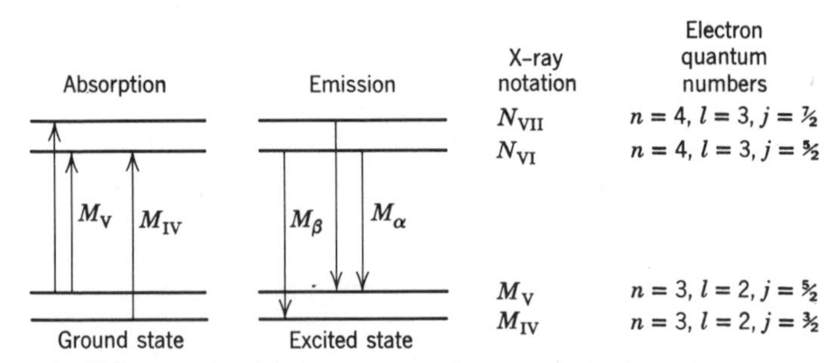

Fig. 9-9. M lines emitted from a rare earth atom (excited state) may be absorbed in a ground state atom by raising a $3d$ electron to the partly filled $4f$ level (after Gilmore and Burr, Ref. 12).

TABLE 9-5

Comparison of X-Ray and Microcombustion Methods for Determination of Carbon and Hydrogen in Petroleum

Petroleum	Carbon, %			Hydrogen, %		
	Comb.[a]	X-ray	Deviation	Comb.[a]	X-ray	Deviation
Wilmington	85.44	85.38	−0.06	11.88	12.56	0.68
West Texas	85.58	85.88	0.30	13.22	12.79	−0.43
Oklahoma City Wilcox	86.46	86.89	0.43	13.45	12.98	−0.47
Bachaquero	84.69	85.51	0.82	12.23	11.50	−0.73
Tatums	85.24	85.69	0.45	12.10	12.53	0.43
Bradford[b]	87.84	86.44	−1.40	13.93	13.46	−0.47

[a] Microcombustion analysis was performed by a commercial microanalytical laboratory.

[b] Combustion results must be too high in this case since C + H = 101.77%.

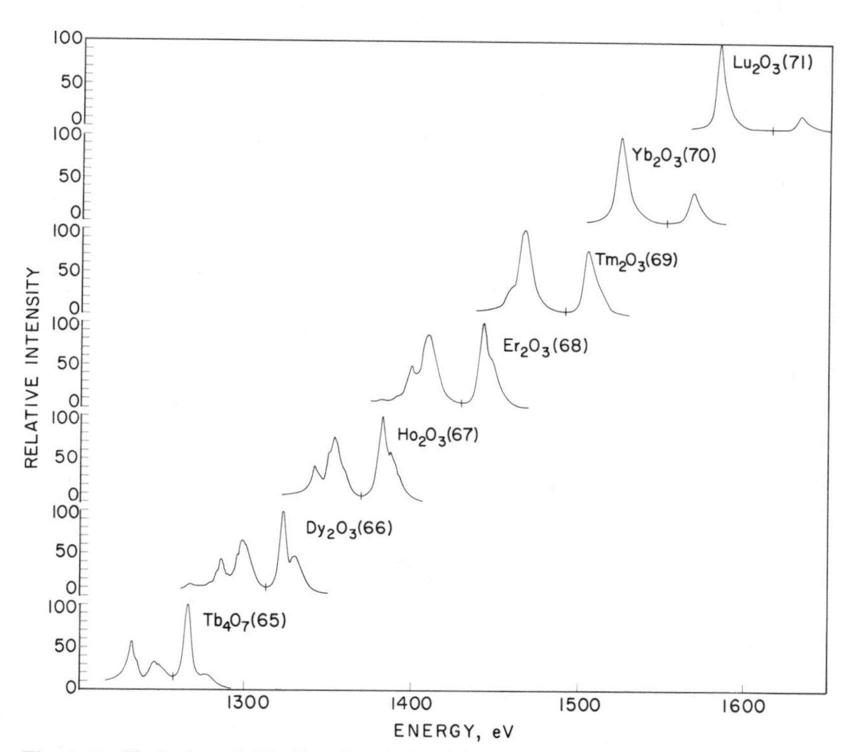

Fig. 9-10. Variation of M_α line shapes for Tb to Lu caused by self-absorption as illustrated in Fig. 9-9 (after Gilmore and Burr, Ref. 12).

pound and used Co K_α as a measure of K content; he also precipitated P as a bismuth compound and used Bi L_α as a measure of P. Rudolph and Nadalin[15] precipitated Cl as AgCl and used Ag K_α as a measure of Cl content. They were able to determine as little as 15 ppm Cl by this method.

References

1. F. Claisse, Can. Dept. Mines and Tech. Sur. Rep. PR 327 (1956).
2. H. F. Beeghly, *J. Electrochem. Soc.*, **97**, 152 (1950).
3. T. N. Rhodin, *Anal. Chem.*, **27**, 1857 (1955).
4. W. J. Campbell, ASTM Spe. Tech. Pub. 349, p. 48, American Soc. for Test. Materials, Philadelphia, Pa., 1963.
5. W. J. Campbell, E. F. Spano, and T. E. Green, *Anal. Chem.*, **38**, 987 (1966).
6. B. L. Henke, *Advances in X-Ray Analysis*, Vol. 8, Plenum Press, New York, 1965, p. 269.
7. W. Ruderman, Isomet Corp., Palisades Park, N. J., private communication.
8. K. F. J. Heinrich, *Advances in X-Ray Analysis*, Vol. 5, Plenum Press, New York, 1962, p. 516.
9. G. A. Anderman, *Anal. Chem.*, **33**, 1689 (1961).
10. J. E. Holliday, *Advances in X-Ray Analysis*, Vol. 9, Plenum Press, New York, 1966, p. 365.
11. D. W. Fischer and W. L. Baun, *J. Appl. Phys.*, **38**, 229 (1967).
12. C. Gilmore and A. Burr, XIII International Spectroscopy Colloquium at Ottawa, Canada, June 1967.
13. C. W. Dwiggens, *Anal. Chem.*, **33**, 67 (1961).
14. G. A. Smith, *Chem. Ind. (London)*, **1963**, 1907.
15. J. S. Rudolph and R. J. Nadalin, *Anal. Chem.*, **36**, 1815 (1964).

CHAPTER 10

ELECTRON PROBE MICROANALYSIS

Electron probe microanalysis[1] is a specialized technique in X-ray spectrochemical analysis. It uses a focused beam of electrons to excite characteristic X-rays in an area as small as one micron in diameter on the sample surface. Thus its purpose is to measure local variations in composition rather than average composition of the whole specimen. This is important in metallurgy, for instance, where it is the composition and distribution of precipitates or the diffusion of constituents near grain boundaries which determine mechanical or chemical properties. It is also important in mineralogy or geology where the localization of compounds is related to the reactions which formed the material and can often be used to predict the presence or absence of other kinds of compounds. In biology, the localization of metal elements in cells, bone, or membranes is a critical factor in body functions and disease.

This chapter gives only a cursory discussion of electron probe analysis for those analysts who wish to know its general capabilities.

10.1. Instrumentation

The electron probe consists of three components as shown schematically in Fig. 10-1:

(*a*) The electron optical system uses an electron gun to generate a beam of electrons and usually two electromagnetic lenses to focus the beam on the sample. Deflection plates or coils allow the focused beam to be swept back and forth in order to scan an area of the surface. The electron optics column and the specimen are in a continuously pumped vacuum system.

(*b*) An optical microscope is used by the operator to select the particular area of the sample to be examined. Photographs may be taken through the microscope or a photoelectric detector may replace the eyepiece and be used to detect cathodoluminescence from certain constituents.

(*c*) Curved-crystal X-ray spectrometers and back-scattered electron detectors are used to record the signals from the specimen for quantitative or qualitative analysis. Scanning electron and/or X-ray images are particularly helpful in delineating the distribution of constituents. Figure 10-2 shows scanning images of an inclusion in steel. Often such

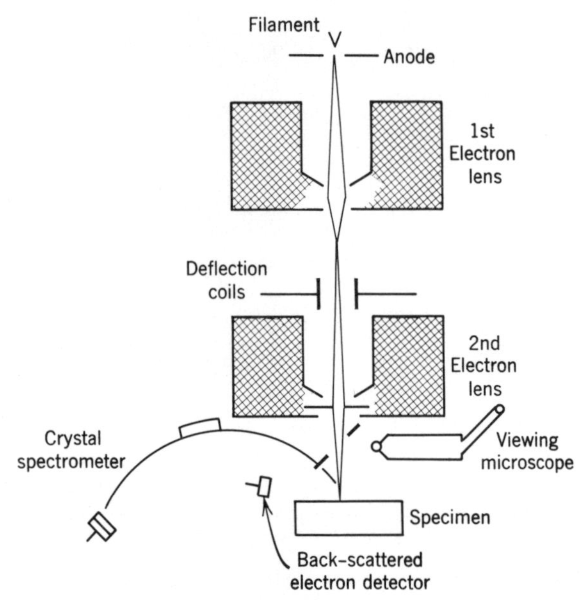

Fig. 10-1. Schematic of electron probe microanalyzer.

qualitative examination is adequate in defining the reactions of interest. When a semiquantitative analysis is desired, the variation in X-ray intensity during scanning may be stored in a multichannel analyzer circuit and printed out as shown in Fig. 10-3. The best quantitative analysis, however, is obtained by positioning the electron beam at the desired spot on the sample and recording the various X-ray lines for sufficient time to obtain adequate statistical precision.

10.2. Specimen Preparation for Hard Materials

The electron beam penetrates only a few microns into the specimen. Therefore, most of the X-ray intensity is generated in this surface layer although some of the characteristic X-ray intensity is also generated at greater depths by the continuum X-ray spectrum excited at the surface by the electrons. Because of the surface nature of the analysis, specimen preparation is more critical than for X-ray fluorescence. For metals, minerals, or similar hard materials, grinding on a graded series of abrasive papers or laps is followed by fine polishing. Diamond dust of the quarter-micron size is perhaps the most suitable material for the final polishing because it cuts hard and soft components of the specimen with equal facility and results in a flat surface without relief.

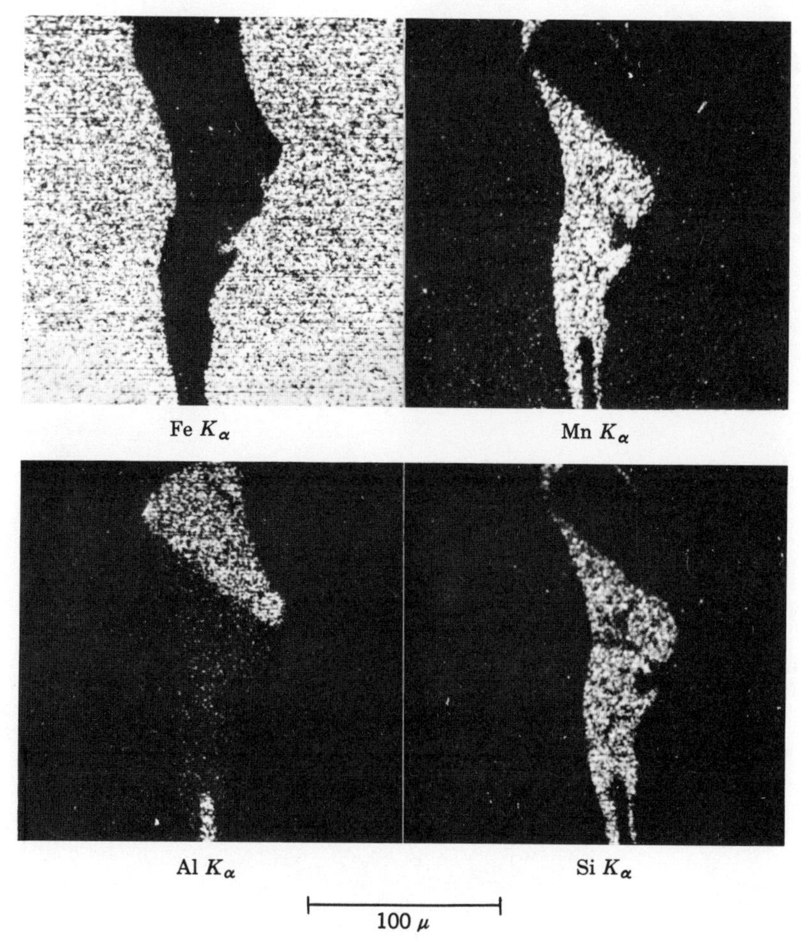

Fig. 10-2. Display pictures of a vitreous inclusion in 1% Cr steel. (Reprinted by permission of CAMECA.)

The metallurgical practice of etching specimens to bring out microscopic detail is not suitable for electron probe examination because chemical etching, by its very nature, selectively removes some constituents and thus changes the surface composition drastically. However, it is feasible to etch the surface in the usual manner, mark areas of interest with a micro-indenter, photograph the area, and then repolish lightly to get below the etched layer but not below the indentation marks. Figure 10-4*a* shows the surface of a Ti–Nb diffusion couple etched and marked; Fig. 10-4*b* shows the same surface after light re-

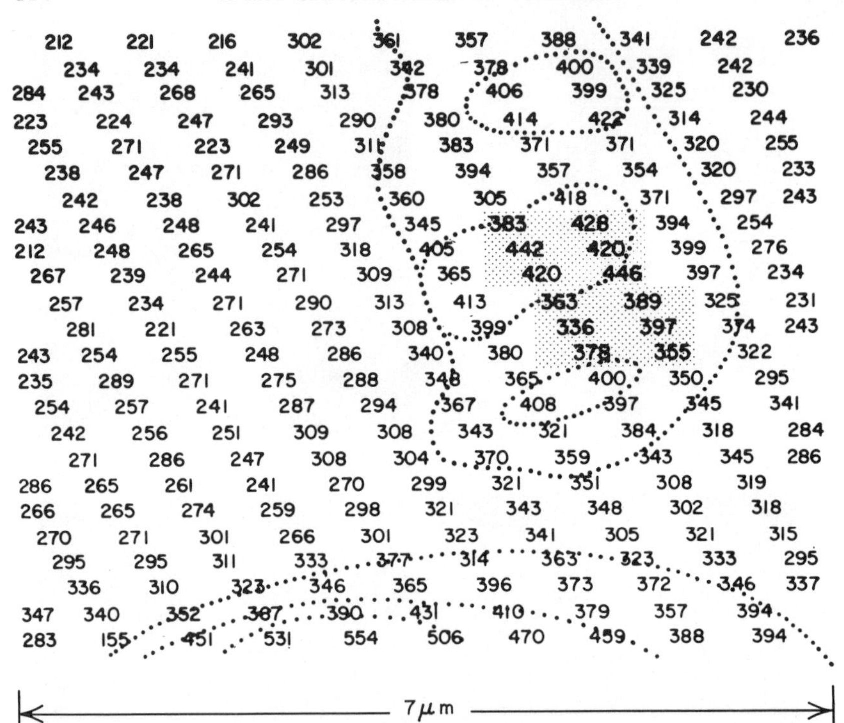

Fig. 10-3. Print-out of intensity topograph to show Fe rich inclusion in Al alloy Numbers are proportional to intensity. Areas outlined manually by connecting equal intensities. Full width of area is 7 μm.

polishing. Using the indentations as starting positions in the electron probe, and with the micrograph as a guide, the operator can easily move to the exact area of interest on the polished surface. Mineral thin sections can be viewed directly by transmission microscopy during the course of electron probe examination; this eliminates some of the problems of area selection.

Samples which are poor electrical conductors must be coated with an evaporated metal film to make them conductors. If this is not done, the surface will charge statically and deflect the electron beam in an erratic manner. The evaporated layer need be only a few hundred angstroms thick and does not alter the excitation or emission of X-rays noticeably. Nevertheless, it is advisable to coat both the unknown and standards (even if the standards are conductors) simultaneously so that any slight alterations of intensity are the same for standards and unknown.

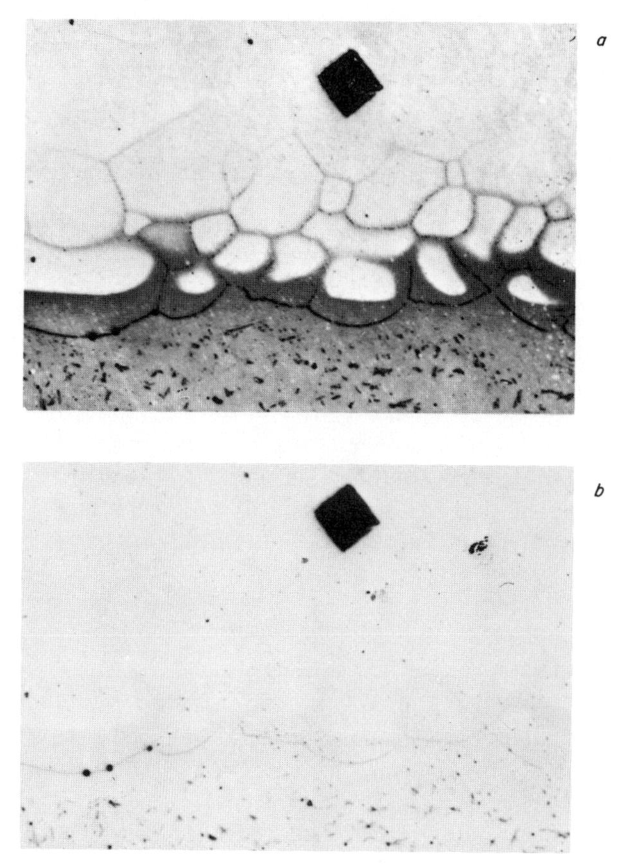

Fig. 10-4. (*a*) Anodized and punch-marked area of grain boundary diffusion in Nb-Ti diffusion zone. (*b*) Same area after anodized layer has been removed by light repolishing.

10.3. Specimen Preparation of Soft Materials

Biological or plant tissues or other soft materials are readily cut into sections with a microtome. They are usually mounted on a support such as a metal substrate, Fig. 10-5, and coated with an evaporated metal film to make the surface conducting. The intense electron beam may damage the specimen and alter its composition if the beam remains too long at one position; thus scanning in one or two directions is the preferred practice. Often the electron beam penetrates through the thin section and the X-ray intensity is reduced to less than that for a bulk specimen. It is desirable that the standards be about the same thickness as the unknown section and similarly mounted so that the intensities can be properly related. Particulate metal compounds suspended in plastic films have been found especially well suited as standards for biological work.[2]

10.4. Particulate Matter; Extraction Replicas

Particles such as dust, sediments, or precipitates must be mounted on a conducting support for electron probe analysis. High-purity aluminum (99.99+%) or evaporated carbon films are suitable substrates. The size and shape of the particles precludes accurate quantitative analysis in most instances but the relative amounts of different elements contained are readily determined.

A special case of particulate matter is the extraction replica which has proved so useful in electron microscopy. It is prepared by polishing an alloy specimen in the usual fashion, Fig. 10-6*a*, and then etching to remove the matrix and free the precipitates or inclusions as shown in Fig 10-6*b*. After the etchant has been drained away carefully to avoid

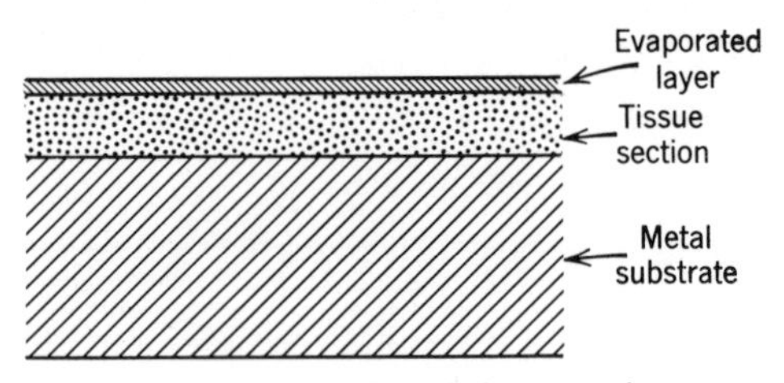

Fig. 10-5. Mounting of tissue section.

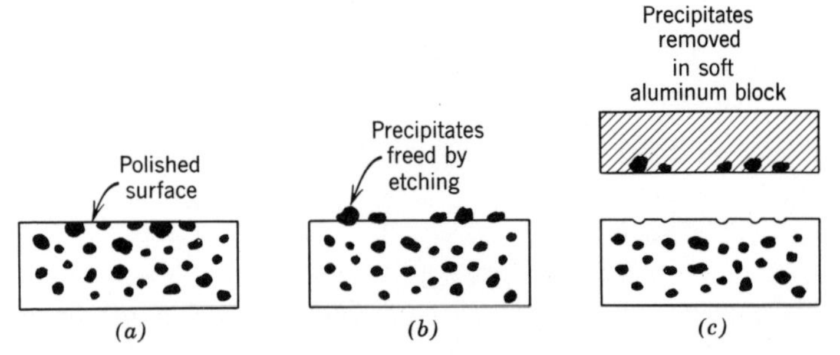

Fig. 10-6. Extraction replica method of sample preparation. (*a*) Alloy containing precipitates is polished. (*b*) Surface is etched to free precipitates from matrix without changing their position. (*c*) Soft aluminum is pressed against etched surface to embed precipitates in the aluminum for electron probe examination.

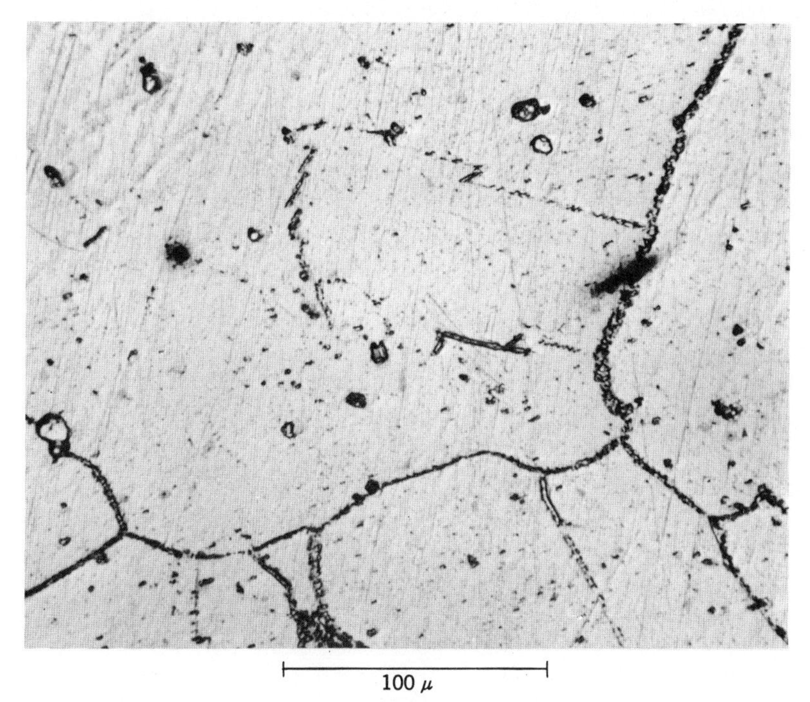

Fig. 10-7. Extraction replica of Ni base alloy with precipitates along brain boundaries.

moving the freed precipitates a block of soft aluminum is pressed against the surface and the precipitates are embedded in the aluminum, Fig. 10-6*c*. This becomes the electron probe specimen. Figure 10-7 shows the extraction replica of an Ni base alloy with precipitates along the grain boundaries. By this means it is possible to measure the composition of precipitates even smaller than the electron-beam size without interference from matrix elements.[3]

10.5. Quantitative Analysis, Standards, and Calculations

The use of comparison standards is much more difficult for electron probe analysis than for X-ray fluorescence because most materials are not homogeneous on a micron-size scale. Defocusing the electron beam to average over a heterogeneous specimen is not satisfactory because the nonlinear nature of the matrix effects (see Sec. 7.2) prevents averaging the intensities from varying compositions. Crystalline compounds may or may not be stoichiometric in composition and should be tested by wet chemistry or X-ray diffraction before they can be used with confidence. Pure elements are satisfactory as end-point standards and are necessary for normalizing all measurements.

Mathematical calculations are widely used for quantitative analysis and give accuracies of a few percent of the amount present in most instances. The format of the calculations is somewhat different than for X-ray fluorescence because the excitation by the electron beam is more difficult to calculate than excitation by primary X-rays (see Sec. 8.5). In addition there is a contribution to the characteristic intensities due to excitation by the continuous X-ray spectrum excited at the surface by the electrons.

The calculation techniques have gone through three stages of development in the history of electron probe analysis:

(*a*) Castaing[4] developed equations which expressed the excitation of characteristic X-rays as a function of depth below the surface[5] and included secondary fluorescence by characteristic X-rays and absorption of X-rays emerging to the surface. His equations have been modified by many workers[6] and additional terms added to account for fluorescence by the continuum[7] and the variation of absolute intensity from element to element as a function of atomic number.[8] These equations are suitable for "pencil and paper" evaluation with the aid of tables of the electron excitation values and the atomic number effect. They necessarily contain approximations which limit their accuracy to a few percent for major constituents.

(*b*) The equations of Castaing and others have been written as computer programs for easier and faster evaluation but with only minor improvements in accuracy.[9] Such programs are available upon request from various laboratories. The accuracy of analysis is about the same as for (*a*) above but computation is facilitated.

(*c*) More elaborate mathematics have been written directly for computer evaluation.[10,11] These avoid many of the approximations required for the earlier equations but the new approaches cannot be solved except by computer. Two of these methods are worthy of mention because they offer the best possibilities for general quantitative analysis in the future.

(*1*) Analytic function method. The total intensity of the characteristic radiation from the element of interest is expressed as a sum of contributions from excitation by electrons, characteristic X-rays, and continuum X-rays.[9] Experimental values are necessary for the electron excitation as a function of depth below the surface, the variation of absolute excitation with atomic number and the spectral distribution of the continuum. These expressions are approximate analytic functions best suited for computer calculation and the treatment is similar to the fundamental parameter method described in Sec. 8.5, that is, relative X-ray intensity from unknown and pure-element standard is the term used. Accuracy is usually slightly better than in (*a*) or (*b*) above.

(*2*) Electron transport method. Electron excitation[1] is calculated directly from the energy loss equations of Bethe, Rose, and Smith.[12] This eliminates the need for experimental values of the electron excitation as a function of depth below the surface and for the variation with atomic number. The other contributions are treated as before. Accuracy is often about one percent composition for major constituents and a few percent of the amount present for minor constituents.

Perhaps the optimum approach is an iterative computer program where the first one or two iterations are accomplished with the faster, but approximate, analytic functions and the final one or two iterations are accomplished with the longer, but more accurate, electron transport calculation. As better values of mass absorption coefficients and fluorescent yields become available, the overall accuracy should approach 1% of the amount present for major constituents.

10.6. Applications

Electron probe analysis has been applied in metallurgy, mineralogy, biology, ceramics, solid state devices, and many other fields. Applications fall under a few general headings.

Identification of Inclusions Precipitates and Localized Enrichments

In the metallurgical field many papers have discussed the composition of precipitates and inclusions in steel and other alloys. In mineralogy and geology, the inclusions in rocks and meteorites have been measured. Glasses and ceramics contain inclusions which may cause discoloration and must be eliminated. Localized zones of enrichment have been found in biological specimens of bone, tissue, and membranes and related to body functions and disease.

Diffusion and Corrosion

One of the major applications of the electron probe in metallurgy has been the measurement of rate and extent of diffusion in binary and ternary alloys. The length of time necessary to diffuse two materials sufficiently for electron probe analysis is only a few hours or days whereas weeks or months might be required if mechanical sections were to be used for other analytical techniques. Formation of intermediate phases in alloys and corrosion products are also important application of diffusion type reactions.

Semiconductor Properties

Transistors and other solid state circuits are very sensitive to mechanical or chemical defects, especially around locations where the leads are attached. The electron probe may be used in the scanning, backscattered electron mode to measure uniformity of the electric field across the solid-state devices while they are in operation and thus to judge their probable operational characteristics. Localized variations in the electric field are easily recognized by sharp boundaries in the backscattered electron image and often may be related to variations in composition from X-ray images.

There are far too many application papers to cite but excellent reviews and bibliographies may be found in the literature.[13]

References

1. L. S. Birks, *Electron Probe Microanalysis*, Interscience, New York, 1963.
2. A. J. Tousimis, ASTM Spec. Tech. Pub. 430, American Soc. for Testing Materials, Philadelphia, Pa., 1968.
3. E. J. Brooks and L. S. Birks, ASTM Spec. Tech. Pub. 245, American Soc. for Testing Materials, Philadelphia, Pa., 1958.
4. R. Castaing, Thesis, Univ. of Paris, 1951.
5. R. Castaing and J. Descamps, *J. Phys. Radium*, **16**, 304 (1955).
6. D. B. Wittry, ASTM Spec. Tech. Pub. 349, p. 128, American Soc. for Test. Materials, Phila., Pa., 1964. P. Duncumb and P. K. Shields, *Brit. J. Appl. Phys.*, **14**, 617 (1963). J. Philibert, (Metaux Corrosion, Ind.) #465, p. 157, #466, p. 216, #469, p. 325 (1964). J. W. Criss and L. S. Birks, *The Electron Microprobe*, Wiley, New York, 1966, p. 217. D. B. Brown and R. E. Ogilvie, *J. Appl. Phys.*, **37**, 4429 (1966).
7. J. Hennoc, F. Maurice, and A. Kirianenko, Centre d'Etude Nucleaires, Saclay, France, Reprint CEA-R 2421, Apr. 1964.
8. P. M. Thomas, Atomic Energy Res. Estab., Harwell, Reprint AERE-R, 4593 (1964).
9. J. D. Brown, *Anal. Chem.*, **38**, 890 (1966).
10. J. W. Criss, ASTM Spec. Tech. Pub. 430, p. 291, American Soc. for Testing Materials, Phila., Pa., 1968.
11. D. B. Brown, D. B. Wittry, and D. F. Kyser, *J. Appl. Phys.*, **40**, 1627 (1969).
12. H. A. Bethe, M. E. Rose, and L. P. Smith, *Proc. Am. Phil. Soc.*, **78**, 573 (1938).
13. W. J. Campbell and J. D. Brown, *Anal. Chem. Review Issue*, **40**, 346R (1968).

APPENDIX 1

SPECTRAL DISTRIBUTION FROM X-RAY TUBES

A.1.1. Experimental Measurements

The spectra from W, Cr, Cu, and Mo X-ray tubes were measured[1] using the arrangement of Fig. A.1-1. Tube voltages were varied from 15 to 50 kV but not all voltages for each target. Radiation from the X-ray tube passes through the slit system, is diffracted by a LiF crystal, and measured with a gas-flow proportional counter using P-10 gas. The slit limits the beam width so that the crystal will intercept the entire beam even at the smallest diffraction angle of 4.5° θ. The tubular collimator merely serves to limit the fanning divergence on the crystal. A knife-edge shields the detector from radiation scattered by the slit edges.

Spectral intensity was measured from the short wavelength limit, which varies from 0.25 to 1.0 Å depending on tube voltage, up to 2.5 or 2.8 Å. Counting circuitry consisted of a multichannel analyzer for visual examination of the complete signal and background at each crystal setting plus a single-channel analyzer set to accept the desired wavelength λ and its escape peak but to exclude higher-order diffractions of λ/n components. Background was measured at the same detector and circuit setting but with the crystal decoupled and rotated out of the diffraction position so it scatters other wavelengths the same as before but does not diffract wavelength λ. Figure A.1-2 shows a typical multichannel display of total signal and background.

In addition to background subtraction, the data were corrected for air-path absorption, detector efficiency, crystal diffraction efficiency, polarization, and change of variable ($\Delta\theta$ to $\Delta\lambda$). Typical spectra were shown in Fig. 3-1 in Chapter 3.

A.1.2. Tabulation of Data

Integrated intensity at each wavelength interval $\Delta\lambda$ and for the characteristic lines are tabulated in Tables A.1-1, -2, -3, -4, and -5.

Contrary to popular belief, the characteristic lines contribute a considerable fraction of the total intensities; 24% for W OEG-50 @ 45 kV (CP), 31% for W OEG-50 @ 50 full wave rectified, 75% for Cr OEG-50 @ 45 kV (CP), and 60% for Cu diffraction tube @ kV (CP).

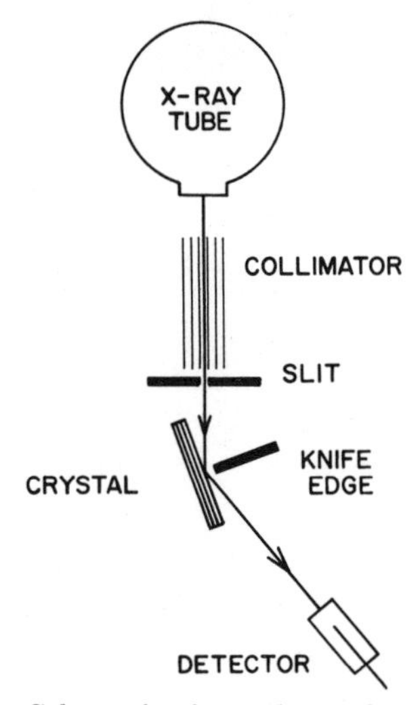

Fig. A.1-1. Schematic of experimental arrangement.

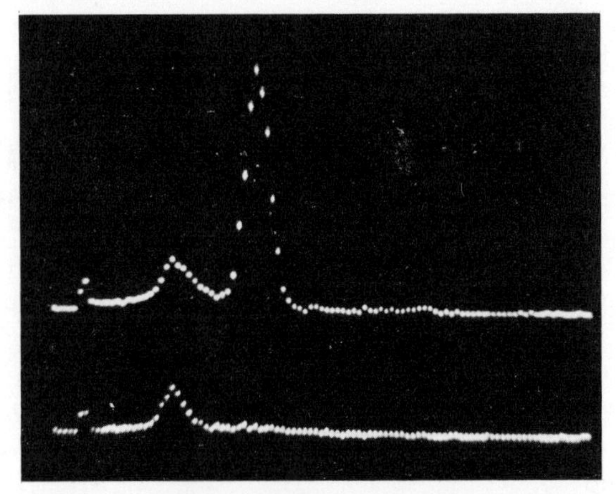

Fig. A.1-2. Multichannel analyzer display of total signal (upper) and background (lower) at 24° 2θ (0.84 Å).

A.1.3. Use of the Table

The tables are intended for use with the fundamental parameter approach to quantitative analysis discussed in Chapter 8. It should be pointed out that any tubes with the same target materials operated at the same voltage, with approximately the same take-off angle (30° for the OEG-50 tubes), and the same window thickness (0.040 in. Be

TABLE A.1-1
Integrated Intensity in the Spectrum of a Tungsten Target Tube at 50 kVP

Continuum

λ[a] (Å)	$I_\lambda \Delta\lambda$	λ[a] (Å)	$I_\lambda \Delta\lambda$	λ[a] (Å)	$I_\lambda \Delta\lambda$	λ[a] (Å)	$I_\lambda \Delta\lambda$
0.27	30.0	0.87	118	1.47	83.7	2.07	31.0
0.29	53.2	0.89	114	1.49	83.1	2.09	29.7
0.31	76.8	0.91	111	1.51	82.3	2.11	28.5
0.33	104	0.93	108	1.53	81.3	2.13	27.4
0.35	124	0.95	105	1.55	80.2	2.15	26.3
0.37	148	0.97	103	1.57	78.9	2.17	25.3
0.39	169	0.99	100	1.59	77.4	2.19	24.3
0.41	184	1.01	97.9	1.61	75.9	2.21	23.4
0.43	196	1.03	95.8	1.63	74.0	2.23	22.6
0.45	202	1.05	93.8	1.65	72.2	2.25	21.8
0.47	205	1.07	91.8	1.67	70.2	2.27	21.1
0.49	206	1.09	90.0	1.69	68.4	2.29	20.4
0.51	205	1.11	88.4	1.71	66.6	2.31	19.6
0.53	202	1.13	87.0	1.73	64.6	2.33	18.8
0.55	195	1.15	85.8	1.75	62.6	2.35	18.0
0.57	187	1.17	84.6	1.77	60.6	2.37	17.3
0.59	180	1.19	83.5	1.79	58.6	2.39	16.8
0.61	173	1.21[d]	84.2	1.81	56.6	2.41	16.3
0.63	167	1.23	90.2	1.83	54.6	2.43	15.7
0.65	163	1.25	89.8	1.85	52.4	2.45	15.2
0.67	159	1.27	89.3	1.87	50.0	2.47	14.9
0.69	154	1.29	88.8	1.89	47.6	2.49	14.7
0.71	150	1.31	88.3	1.91	45.4	2.51	14.5
0.73	146	1.33	87.8	1.93	43.3	2.53	14.3
0.75	141	1.35	87.4	1.95	41.2	2.55	14.1
0.77	137	1.37	86.9	1.97	39.2	2.57	13.8
0.79	133	1.39	86.4	1.99	37.3	2.59	13.4
0.81	129	1.41	85.8	2.01	35.6		
0.83	125	1.43	85.1	2.03	34.0		
0.85	121	1.45	84.4	2.05	32.4		

Characteristic Lines[c]

	λ[b] (Å)	$I_\lambda \Delta\lambda$[c]
$L_{\gamma_{2+3}}$	1.06460	68.8
L_{γ_1}	1.09855	155
L_{β_2}	1.24460	532
L_{β_1}	1.281809	1464
L_{α_1}	1.47639	2072
L_l	1.6782	68.8

[a] λ for continum is the middle of the $\Delta\lambda$ interval.

[b] λ for lines is from Bearden (weighted average for $L_{\gamma_{2+3}}$.

[c] $\Delta\lambda$ for lines is natural line breadth (from Blohkin, *Physics of X-rays*, AEC translation 4502).

[d] L_{III} edge occurs at 1.216 Å. ($I_\lambda \Delta\lambda$ is 66.1 from 1.200 to 1.216 and 18.1 from 1.216 to 1.220 Å).

TABLE A.1-2

W Target OEG-50 45 kV (CP) $\Delta\lambda = 0.02$ Å

Continuum

λ[a] (Å)	$I_\lambda\Delta\lambda$	λ[a] (Å)	$I_\lambda\Delta\lambda$	λ[a] (Å)	$I_\lambda\Delta\lambda$	λ[a] (Å)	$I_\lambda\Delta\lambda$
0.29	15.5	0.87	44.0	1.45	30.9	2.03	10.4
0.31	36.6	0.89	42.9	1.47	30.3	2.05	10.0
0.33	56.8	0.91	42.0	1.49	29.7	2.07	9.7
0.35	76.6	0.93	41.2	1.51	29.0	2.09	9.4
0.37	96.2	0.95	40.4	1.53	28.3	2.11	9.0
0.39	111.1	0.97	39.6	1.55	27.6	2.13	8.6
0.41	116.4	0.99	38.8	1.57	26.8	2.15	8.2
0.43	114.6	1.01	38.1	1.59	26.0	2.17	7.8
0.45	109.9	1.03	37.4	1.61	25.2	2.19	7.5
0.47	104.5	1.05	36.7	1.63	24.3	2.21	7.3
0.49	99.1	1.07	36.1	1.65	23.2	2.23	7.0
0.51	93.9	1.09	35.6	1.67	22.1	2.25	6.7
0.53	89.0	1.11	35.1	1.69	21.1	2.27	6.4
0.55	84.4	1.13	34.5	1.71	20.1	2.29	6.1
0.57	80.3	1.15	33.9	1.73	19.2	2.31	5.9
0.59	76.6	1.17	33.4	1.75	18.3	2.33	5.7
0.61	73.3	1.19	33.0	1.77	17.5	2.35	5.4
0.63	70.1	1.21[b]	33.4	1.79	16.8	2.37	5.1
0.65	67.0	1.23	36.2	1.81	16.1	2.39	4.9
0.67	64.1	1.25	35.8	1.83	15.5	2.41	4.7
0.69	61.2	1.27	35.4	1.85	14.9	2.43	4.5
0.71	58.6	1.29	35.0	1.87	14.3	2.45	4.2
0.73	56.2	1.31	34.6	1.89	13.7	2.47	3.9
0.75	53.9	1.33	34.1	1.91	13.1	2.49	3.7
0.77	51.8	1.35	33.5	1.93	12.6	2.51	3.5
0.79	49.9	1.37	33.0	1.95	12.2	2.53	3.3
0.81	48.2	1.39	32.5	1.97	11.7	2.55	3.1
0.83	46.6	1.41	32.0	1.99	11.2	2.57	2.9
0.85	45.2	1.43	31.5	2.01	10.8	2.59	2.7

Characteristic Lines

	λ[c] (Å)	$I_\lambda\Delta\lambda$
L_{γ_3}	1.06200	2.6
L_{γ_2}	1.06806	10.4
L_{γ_1}	1.09855	27.8
L_{β_2}	1.24460	180
L_{β_1}	1.281809	407
L_α	1.47639	592
L_l	1.6782	13.8

[a] λ for continuum is the middle of the $\Delta\lambda$ interval.

[b] L_{III} edge occurs at 1.216 Å ($I_\lambda\Delta\lambda$ is 26.1 from 1.200 to 1.216 and 7.3 from 1.216 to 1.220 Å).

[c] λ for lines is from Bearden. $\Delta\lambda$ for lines is natural line breadth (from Blohkin, *Physics of X-Rays*, AEC Translation 4502).

TABLE A.1-3

Cr Target OEG-50 45 kV (CP) $\Delta\lambda = 0.02$ Å

Continuum

λ[a] (Å)	$I_\lambda \Delta\lambda$	λ[a] (Å)	$I_\lambda \Delta\lambda$	λ[a] (Å)	$I_\lambda \Delta\lambda$	λ[a] (Å)	$I_\lambda \Delta\lambda$
0.29	3.00	0.97	12.4	1.65	5.12	2.33	6.13
0.31	6.60	0.99	12.1	1.67	5.00	2.35	5.94
0.33	8.80	1.01	11.8	1.69	4.88	2.37	5.74
0.35	9.94	1.03	11.4	1.71	4.77	2.39	5.55
0.37	11.0	1.05	11.1	1.73	4.66	2.41	5.37
0.39	12.0	1.07	10.9	1.75	4.55	2.43	5.20
0.41	12.9	1.09	10.6	1.77	4.44	2.45	5.03
0.43	13.5	1.11	10.3	1.79	4.33	2.47	4.86
0.45	13.9	1.13	10.0	1.81	4.22	2.49	4.70
0.47	13.8	1.15	9.78	1.83	4.11	2.51	4.55
0.49	13.8	1.17	9.51	1.85	4.01	2.53	4.42
0.51	13.8	1.19	9.25	1.87	3.90	2.55	4.28
0.53	13.8	1.21	9.00	1.89	3.79	2.57	4.15
0.55	13.8	1.23	8.76	1.91	3.69	2.59	4.02
0.57	13.8	1.25	8.52	1.93	3.59	2.61	3.92
0.59	13.8	1.27	8.29	1.95	3.49	2.63	3.83
0.61	13.8	1.29	8.07	1.97	3.39	2.65	3.73
0.63	13.8	1.31	7.84	1.99	3.29	2.67	3.64
0.65	13.8	1.33	7.62	2.01	3.19	2.69	3.56
0.67	13.9	1.35	7.42	2.03	3.09	2.71	3.47
0.69	13.9	1.37	7.22	2.05	2.98	2.73	3.38
0.71	14.0	1.39	7.02	2.07[b]	5.92	2.75	3.31
0.73	14.0	1.41	6.83	2.09	8.77	2.77	3.24
0.75	14.0	1.43	6.65	2.11	8.54	2.79	3.16
0.77	14.0	1.45	6.48	2.13	8.30	2.81	3.09
0.79	14.0	1.47	6.32	2.15	8.07	2.83	3.02
0.81	13.9	1.49	6.16	2.17	7.84	2.85	2.95
0.83	13.8	1.51	6.02	2.19	7.61	2.87	2.89
0.85	13.7	1.53	5.89	2.21	7.39	2.89	2.82
0.87	13.6	1.55	5.75	2.23	7.17	2.91	2.75
0.89	13.4	1.57	5.62	2.25	6.96	2.93	2.68
0.91	13.2	1.59	5.50	2.27	6.74	2.95	2.61
0.93	13.0	1.61	5.38	2.29	6.53	2.97	2.55
0.95	12.7	1.63	5.25	2.31	6.32	2.99	2.49

Characteristic Lines

	λ[c] (Å)	$I_\lambda \Delta\lambda$
K_β	2.08478	415
K_α	2.29100	2511

[a] λ for continuum is the middle of $\Delta\lambda$ interval.

[b] K edge occurs at 2.070 Å ($I_\lambda \Delta\lambda$ is 1.45 from 2.06 to 2.07 and 4.47 from 2.07 to 2.08 Å).

[c] λ for lines is from Bearden (weighted average for $K_{\alpha_{1+2}}$) and $\Delta\lambda$ for lines is natural line breadth (from Blohkin, *Physics of X-Rays*, AEC Translation 4502).

TABLE A.1-4

Cu Target OEG-50 45 kV (CP) $\Delta\lambda = 0.02$ Å

Continuum

λ[a] (Å)	$I_\lambda\Delta\lambda$	λ[a] (Å)	$I_\lambda\Delta\lambda$	λ[a] (Å)	$I_\lambda\Delta\lambda$	λ[a] (Å)	$I_\lambda\Delta\lambda$
0.29	1.65	0.87	11.8	1.45	8.10	2.03	3.08
0.31	4.09	0.89	11.6	1.47	7.89	2.05	2.96
0.33	6.47	0.91	11.4	1.49	7.68	2.07	2.85
0.35	8.88	0.93	11.2	1.51	7.48	2.09	2.75
0.37	11.2	0.95	11.0	1.53	7.28	2.11	2.65
0.39	13.2	0.97	10.7	1.55	7.08	2.13	2.56
0.41	14.3	0.99	10.4	1.57	6.88	2.15	2.47
0.43	14.6	1.01	10.0	1.59	6.68	2.17	2.37
0.45	14.7	1.03	9.55	1.61	6.48	2.19	2.28
0.47	14.7	1.05	9.08	1.63	6.28	2.21	2.20
0.49	14.7	1.07	8.62	1.65	6.09	2.23	2.13
0.51	14.7	1.09	8.20	1.67	5.90	2.25	2.06
0.53	14.7	1.11	7.78	1.69	5.71	2.27	1.97
0.55	14.6	1.13	7.36	1.71	5.53	2.29	1.89
0.57	14.6	1.15	7.00	1.73	5.35	2.31	1.83
0.59	14.5	1.17	6.70	1.75	5.17	2.33	1.76
0.61	14.4	1.19	6.45	1.77	5.00	2.35	1.69
0.63	14.4	1.21	6.25	1.79	4.83	2.37	1.63
0.65	14.3	1.23	6.08	1.81	4.66	2.39	1.56
0.67	14.2	1.25	5.94	1.83	4.50	2.41	1.49
0.69	14.1	1.27	5.82	1.85	4.34	2.43	1.43
0.71	13.9	1.29	5.72	1.87	4.19	2.45	1.36
0.73	13.7	1.31	5.65	1.89	4.04	2.47	1.29
0.75	13.4	1.33	5.60	1.91	3.89	2.49	1.24
0.77	13.1	1.35	5.56	1.93	3.75	2.51	1.19
0.79	12.8	1.37	5.51	1.95	3.61	2.53	1.13
0.81	12.4	1.39[b]	8.73	1.97	3.47	2.55	1.07
0.83	12.2	1.41	8.51	1.99	3.33	2.57	1.01
0.85	12.0	1.43	8.30	2.01	3.20	2.59	0.85

Characteristic Lines

	λ[c] (Å)	$I_\lambda\Delta\lambda$
K_β	1.392218	235
K_α	1.541841	1290

[a] λ for continuum is the middle of $\Delta\lambda$ interval.

[b] K edge occurs at 1.380 Å.

[c] λ for lines is from Bearden. $\Delta\lambda$ for lines is natural line breadth (from Blohkin, *Physics of X-Rays*, AEC Translation 4502).

TABLE A.1-5

Mo Target OEG-50 45 kV (CP) $\Delta\lambda = 0.02$ Å

Continuum

λ[a] (Å)	$I_\lambda \Delta\lambda$	λ[a] (Å)	$I_\lambda \Delta\lambda$	λ[a] (Å)	$I_\lambda \Delta\lambda$	λ[a] (Å)	$I_\lambda \Delta\lambda$
0.29	0.37	0.87	4.06	1.45	2.29	2.03	0.96
0.31	0.88	0.89	4.06	1.47	2.23	2.05	0.92
0.33	1.34	0.91	4.06	1.49	2.17	2.07	0.88
0.35	1.77	0.93	4.05	1.51	2.11	2.09	0.84
0.37	2.18	0.95	4.04	1.53	2.06	2.11	0.81
0.39	2.58	0.97	4.02	1.55	2.01	2.13	0.78
0.41	2.94	0.99	3.98	1.57	1.95	2.15	0.74
0.43	3.18	1.01	3.93	1.59	1.90	2.17	0.71
0.45	3.32	1.03	3.87	1.61	1.85	2.19	0.68
0.47	3.43	1.05	3.81	1.63	1.80	2.21	0.65
0.49	3.53	1.07	3.74	1.65	1.76	2.23	0.63
0.51	3.62	1.09	3.65	1.67	1.71	2.25	0.61
0.53	3.68	1.11	3.56	1.69	1.66	2.27	0.58
0.55	3.70	1.13	3.48	1.71	1.61	2.29	0.55
0.57	3.69	1.15	3.40	1.73	1.56	2.31	0.53
0.59	3.67	1.17	3.32	1.75	1.52	2.33	0.51
0.61	3.64	1.19	3.24	1.77	1.48	2.35	0.49
0.63[b]	4.96	1.21	3.16	1.79	1.44	2.37	0.47
0.65	4.86	1.23	3.08	1.81	1.40	2.39	0.45
0.67	4.78	1.25	3.00	1.83	1.36	2.41	0.43
0.69	4.67	1.27	2.92	1.85	1.32	2.43	0.41
0.71	4.57	1.29	2.84	1.87	1.28	2.45	0.39
0.73	4.48	1.31	2.77	1.89	1.24	2.47	0.37
0.75	4.40	1.33	2.70	1.91	1.20	2.49	0.36
0.77	4.32	1.35	2.63	1.93	1.16	2.51	0.35
0.79	4.24	1.37	2.56	1.95	1.12	2.53	0.33
0.81	4.16	1.39	2.49	1.97	1.08	2.55	0.31
0.83	4.09	1.41	2.43	1.99	1.04	2.57	0.30
0.85	4.06	1.43	2.36	2.01	1.00	2.59	0.29

Characteristic Lines

	λ[c] (Å)	$I_\lambda \Delta\lambda$
K_β	0.632288	11.8
K_α	0.710730	66.8

[a] λ for continuum is the middle of $\Delta\lambda$ interval.

[b] K edge occurs at 0.620 Å.

[c] λ for lines is from Bearden. $\Delta\lambda$ for lines is natural line breadth (from Blohkin, *Physics of X-Rays*, AEC Translation 4502).

for the W, Cu, and Mo; 0.010 in. for Cr) should give the same spectral distributions as those shown. It is easy to correct for different window thickness by the usual absorption equation. Correction for different take-off angles is not as easy but only negligible errors should result for a difference of $\pm 5°$.

Reference

1. J. V. Gilfrich and L. S. Birks, *Anal. Chem.*, **40**, 1077 (1968).

APPENDIX 2

USEFUL ANALYZER CRYSTALS

Crystal	Planes	2d spacing (Å)	Remarks
LiF	420	1.79	Best dispersion for short wavelength
LiF	220	2.84	Better intensity than topaz
LiF	200	4.02	Best intensity
NaCl	200	5.62	
KCl	200	6.28	
Si	111	6.27	No second-order diffraction
Quartz	1011	6.70	
Carbon	002	6.70	Strong intensity, broad lines
Quartz	1010	8.50	
Penta-erythritol	002	8.76	Good intensity but deteriorates gradually
EDDT*	020	8.80	
ADP*	110	10.64	
Mica	002	19.8	Strong higher-order reflections
KAP*	001	26.6	Watch out for fluorescence of potassium
OHM*		63.5	Not readily available
LSD*	Not true crystal	100.2	Multilayer films

The Bragg angle θ is obtained from the equation

$$n\lambda = 2d \sin \theta$$

where λ is the wavelength in angstroms and n is in the order of diffraction.

For an excellent tabulation of wavelengths of all the elements and series lines see J. A. Bearden, *X-Ray Wavelengths*, U.S. Atomic Energy Commission Division of Tech. Information Reprint NYO-10586 (1964). See also J. A. Bearden, *Rev. Mod. Phys.*, **31**, 1 (1967), and J. A. Bearden and A. F. Burr, *Rev. Mod. Phys.*, **31**, 49 (1967).

* EDDT, ethylene diamine d-tartrate; ADP, ammonium dihydrogen phosphate; KAP potassium acid phthalate; OHM, octadecyl hydrogen maleate; LSD, lead stearate decanoate.

APPENDIX 3

AVERAGE VALUES OF FLUORESCENT YIELDS

The values given in Tables A.3-1 and A.3-2 were determined from the curves plotted by Fink et al.[1]

TABLE A.3-1

Average K Fluorescent Yields

Z	Element	ω_K	Z	Element	ω_K	Z	Element	ω_K
6	C	0.0009	24	Cr	0.26	41	Nb	0.755
7	N	0.0015	25	Mn	0.285	42	Mo	0.77
8	O	0.0022	26	Fe	0.32	43	Tc	0.785
10	Ne	0.0100	27	Co	0.345	44	Ru	0.80
11	Na	0.020	28	Ni	0.375	45	Rh	0.81
12	Mg	0.030	29	Cu	0.41	46	Pd	0.82
13	Al	0.040	30	Zn	0.435	47	Ag	0.83
14	Si	0.055	31	Ga	0.47	48	Cd	0.84
15	P	0.070	32	Ge	0.50	49	In	0.85
16	S	0.090	33	As	0.53	50	Sn	0.86
17	Cl	0.105	34	Se	0.565	51	Sb	0.87
18	Ar	0.125	35	Br	0.60	52	Te	0.875
19	K	0.140	36	Kr	0.635	53	I	0.88
20	Sc	0.165	37	Rb	0.665	54	Xe	0.89
21	Ca	0.190	38	Sr	0.685	55	Cs	0.895
22	Ti	0.220	39	Y	0.71	56	Ba	0.90
23	V	0.240	40	Zr	0.73			

TABLE A.3-2

Average L Fluorescent Yields[a]

Z	Element	ω_L	Z	Element	ω_L	Z	Element	ω_L
40	Zr	0.057	59	Pr	0.168	77	Ir	0.340
41	Nb	0.061	60	Nd	0.173	78	Pt	0.353
42	Mo	0.067	61	Pm	0.178	79	Au	0.363
43	Tc	0.073	62	Sm	0.183	80	Hg	0.373
44	Ru	0.080	63	Eu	0.190	81	Tl	0.382
45	Rh	0.085	64	Gd	0.196	82	Pb	0.391
46	Pd	0.091	65	Tb	0.203	83	Bi	0.399
47	Ag	0.096	66	Dy	0.208	84	Po	0.405
48	Cd	0.103	67	Ho	0.213	85	At	0.410
49	In	0.109	68	Er	0.220	86	Ru	0.417
50	Sn	0.115	69	Tm	0.227	87	Fr	0.423
51	Sb	0.120	70	Yb	0.237	88	Ra	0.428
52	Te	0.125	71	Lu	0.252	89	Ac	0.431
53	I	0.131	72	Hf	0.268	90	Th	0.436
54	Xe	0.138	73	Ta	0.285	91	Pa	0.440
55	Cs	0.145	74	W	0.302	92	U	0.443
56	Ba	0.152	75	Re	0.315	93	Np	0.445
57	La	0.157	76	Os	0.327	94	Pu	0.448
58	Ce	0.163						

[a] The values are given to 2 or 3 places to distinguish adjacent elements but any individual value is only accurate to perhaps 5 to 10%.

Reference

1. R. W. Fink, R. C. Jopson, H. Mark, and C. D. Swift, *Rev. Mod. Phys.*, **38**, 513 (1966).

APPENDIX 4

STEPS IN SELECTING COMPONENTS AND CONDITIONS AND IN PERFORMING QUANTITATIVE ANALYSIS

In this Appendix we shall outline the steps taken by an analyst in selecting the optimum components and conditions for a specific type of specimen and the sequence of measurements and data treatment for quantitative analysis. The example is fictitious but illustrates all of the facets discussed in the body of the book. The problem: analyze alloys of the following nominal composition

Si	Al	Fe	Ni	Cu
3%	65%	5%	25%	2%

A.4.1. Specimen Preparation

We would prefer to analyze the solid alloy because solution is time consuming and expensive. If solution is to be used, the borax bead technique would be suitable for Fe, Ni, and Cu but not very suitable for Si or Al because of their low atomic numbers.

To prepare solid samples, they should be ground on a series of abrasive papers down to 600 Aloxite which has a particle size of about 25–50 μ. Even this fine a grinding is not suitable for the analysis of Si unless the grinding marks are oriented parallel to the incident and emergent X-rays.

Microscopy or electron probe examination would be useful to insure that the samples are generally homogeneous. If they are not homogeneous, one must be sure that any comparison standards have similar heterogeneity. If the heterogeneity should be in the form of precipitates, it is important to know which elements are precipitated. Satisfactory analysis of heterogeneous samples in the long-wavelength region may not be possible.

A.4.2. Equipment

Choice of X-Ray Tube

A standard tungsten target tube would be suitable for Fe, Ni, and Cu but a thin window Cr target tube would be required for Si and Al. It would be simplest to use the Cr target tube for all of the elements.

Spectrometer and Crystal

A vacuum spectrometer is required for the Si and Al and may as well be used for the other elements too. A LiF crystal should be used for Fe, Ni, and Cu. A long spacing crystal such as KAP, EDDT, or pentaerythritol should be used for Si and Al.

Detector

A sealed Xe proportional counter or scintillation counter, or a solid state detector should be used for Fe, Ni, and Cu. A flow proportional counter with 90% Ar–10% methane should be used for Si and Al.

A.4.3. Measurements

Expected Counting Rates

As a very rough rule-of-thumb, one can expect about 5000 cps for 1% composition of elements in the neighborhood of Fe using a W target X-ray tube operated at 50 kV, 50 mA provided there is not strong matrix absorption. For Al and Si, one can expect about 500 cps for each 1% composition using a Cr target X-ray tube operated at about 30 kV, 40 mA with the same restriction on matrix absorption. For the hypothetical alloy considered, Table A.4-1 shows the absorption coefficients of interest.

TABLE A.4-1

Radiation (All K_α)

Absorber	Al 8.32	Si 7.11	Fe 1.93	Ni 1.65	Cu 1.54
Al	300	3300	92	59	47
Si	440	295	117	76	63
Fe	3200	2150	76	400	350
Ni	3150	2280	94	59	48
Cu	3420	2500	100	64	54

For the Al radiation, all of the other elements except Si have very strong absorption and there are nearly 35% total of the other elements. As a very rough rule-of-thumb, we may say that when there is strong matrix absorption the intensity should be about

$$\frac{(\text{count rate}/\%) \times (\% \text{ of element})}{(\text{increased absorption}) \times (\text{fractional composition of matrix})}$$

Substituting we obtain for Al K_α

$$\frac{500 \times 65}{10 \times 0.35} \approx 9000 \text{ cps}$$

For the Si radiation, by the same treatment, we would expect

$$\frac{500 \times 3}{10 \times 0.97} \approx 150 \text{ cps}$$

For Fe radiation, there will be only slightly increased absorption in the matrix elements and some enhancement by Ni and Cu. Roughly we would expect

$$5000 \times 5 \approx 25000 \text{ cps}$$

For Ni radiation, only Fe has a strong absorption and there is only 5% of Fe present. We would expect

$$5000 \times 25 \approx 125000 \text{ cps}$$

For Cu radiation, again only the Fe has a strong absorption. We would expect

$$5000 \times 2 \approx 10000 \text{ cps}$$

Expected Resolution

The resolution of LiF is more than adequate to separate the K_α lines of Fe, Ni, and Cu from each other and from the K_β lines of each other as well. The resolution of KAP, EDDT, or pentaerythritol is more than adequate to separate the K_α line of Si from the K_β line of Al. Pulse amplitude discrimination will separate all higher order diffraction of Fe, Ni, and Cu for first order diffraction of Al and Si. Without pulse amplitude discrimination the 5th order of Ni K_α overlaps the 1st order of Al K_α.

Fluorescence of Crystal

There will be some fluorescence of potassium in KAP by the Fe, Ni, and Cu radiation. This may lend to an undesirably high background especially for the low Si signal. A collimator between the crystal and detector will reduce such background appreciably. It may be necessary to use EDDT or pentaerythritol if the background is intolerable.

Tube Operation

For Fe, Ni, and Cu the tube should be operated at 50 keV and the full current. For Al and Si it should be operated at perhaps 30 keV and full current to achieve the optimum balance of full line intensity and minimum background intensity.

A.4.4. Evaluation of Results

Precision for Individual Measurements

For Al, Fe, Ni, and Cu it should be easy to achieve $N_L \approx 100{,}000$ in reasonable counting times. The background should be less than 10% of the line so the overall precision should be about 0.4% of the amount present. For the Si the count rate is low and it would be difficult to achieve an N_L of more than about 10,000 with a background expected to be perhaps 50% of the line in the same counting interval. This would lead to an expected precision of

$$100 \times (15000)^{1/2}/5000 \approx 2.5\% \text{ of the amount present}$$

Accuracy of Analysis

If analysis is done by calibration standards with similar alloys, a ratio method must be used to determine composition. The variance $\sigma^2{}_{A/B}$ for a ratio is just the sum of $\sigma_A{}^2 + \sigma_B{}^2$. The variance in measuring the standards would be about the same as for the unknown making the expected accuracy for Al, Fe, Ni, and Cu about $[(0.4)^2 + (0.4)^2]^{1/2} \approx 0.6\%$ of the amount present. For Si the expected accuracy would be $[(2.5)^2 + (2.5)^2]^{1/2} \approx 3.5\%$ of the amount present. Neither of the above estimates includes possible errors in the standard analysis itself which may well be beyond the limits of the estimates due to X-ray measurements and may raise the errors to more like 1% of the amount present for Al, Fe, Ni, and Cu.

If analysis is done by mathematical methods, the results of Chapter 8 show errors of 0.5–5% of the amount present depending on element and composition but there is no increase in errors due to standards because the only standards are the pure elements themselves.

A.4.5. Comments

The discussion in this Appendix is intended to show that the analyst may do a rapid pencil-and-paper treatment of a given problem merely by making use of the principles outlined in the body of the book. Thus he may decide on the optimum specimen preparation, the type of equipment necessary and the expected accuracy before ever making a single measurement. Time spent in this advance planning is more than repaid by saving haphazard tests to determine the same factors. It should be pointed out, however, that even the best advance planning cannot foresee all possible pitfalls and the analyst should not be discouraged if his predictions are not fulfilled exactly. At least, hoepfully, he will be able to recognize deviations from his predictions and the possible causes.

AUTHOR INDEX

Note: Most of the authors listed are not mentioned by name in the text, but their work is referred to and the exact references given at the ends of the chapters.

SUBJECT INDEX